Safety and Risk in Society

Critical Issues and Analysis in Fire Protection and Prevention

SAFETY AND RISK IN SOCIETY

Additional books and e-books in this series can be found on Nova's website under the Series tab.

SAFETY AND RISK IN SOCIETY

CRITICAL ISSUES AND ANALYSIS IN FIRE PROTECTION AND PREVENTION

AXEL CABLÉ

AND

BRADY MANESCAU

EDITORS

nova science publishers
New York

Library of Congress Cataloging-in-Publication Data

ISBN: 978-1-53618-738-0
Names: Cablé, Axel, 1986- editor. | Manescau, Brady, editor.
Title: Critical issues and analysis in fire protection and prevention / Axel Cablé (editor), General Manager, EXEC, Paris, France, R&D Manager, Consulting in Fire and Industrial Safety; Brady Manescau (editor).
Description: New York : Nova Science Publishers, [2020] | Series: Safety and risk in society | Includes bibliographical references and index. |
Identifiers: LCCN 2020044023 (print) | LCCN 2020044024 (ebook) | ISBN 9781536187380 (hardcover) | ISBN 9781536187960 (adobe pdf)
Subjects: LCSH: Fire protection engineering. | Fire prevention.
Classification: LCC TH9115 .C75 2020 (print) | LCC TH9115 (ebook) | DDC 628.9/22--dc23
LC record available at https://lccn.loc.gov/2020044023
LC ebook record available at https://lccn.loc.gov/2020044024

Published by Nova Science Publishers, Inc. † New York

CONTENTS

PREFACE

With an estimated 135 000 reported fire incidents occurring each day worldwide (CTIF World Fire Statistics Center, 2020), fire safety is a major societal and safety issue. Fires have an enormous impact on people's lives and wellbeing. They cause massive amounts of air pollution, increase carbon emissions, and are responsible for the loss of invaluable natural and cultural heritage.

In the current context of climate change, it is hence crucial to develop appropriate preventive and protective measures against fire. This book is a compilation of studies and advances on the current state of research related to critical issues and analysis in the field of fire safety, with chapter contributions from various countries and research institutions worldwide (Australia, Chile, France, India and USA). It aims to provide a broad picture of currently faced challenges and potential solutions from passive and active protective measures, to modelling and experimentations related to compartment and wildland fires.

- Chapter 1 and 2 address the issues of smoke generation in compartment fires, with, on the one hand, the risk of ignition of unburnt gases and, and on the other hand, the need for visibility through smoke for evacuation.

- Chapter 3 and 4 are dedicated to the fire resistance and pyrolysis study of innovative composite materials and bio-based flame retardants, notably for vehicle and aircraft applications.
- Chapter 5 discusses the challenges related to the fire protection of built cultural heritage and evaluate the performance of an innovative sprinkler fire suppression system.
- Chapter 6 addresses the issue of forest fires and provides a methodology to improve the emergency response time at urban-forest interfaces.
- Chapter 7 opens up the discussion and provides potential solutions to help reduce the occurrence of wildfires and increase community safety in the context of climate change by involving renewable energies and local fire stations.

In: Critical Issues and Analysis … ISBN: 978-1-53618-738-0
Editors: A. Cablé and B. Manescau

Chapter 1

EFFECT OF THE LEVEL OF INSULATION OF AN ENCLOSURE ON THE RISK OF IGNITION OF SMOKE RICH IN UNBURNT GASES IN A VENTILATION SYSTEM: AN EXPERIMENTAL STUDY

Brady Manescau[1,*], Hui-Ying Wang[2], Jean-Pierre Garo[2] and Bruno Coudour[1]

[1]INSA Centre Val de Loire, University of Orléans, PRISME EA 4229, Bourges, France

[2]Institut PPRIME, UPR CNRS 3346, Poitiers, France

[*] Corresponding Author's Email: brady.manescau@insa-cvl.fr.

ABSTRACT

This chapter presents the effect of the insulation level of the walls of an enclosure on the production of unburnt gases in the case of an under-ventilated fire. Depending on the amount of unburnt gases, it is possible to assess the risk of inflammation of the gases considering their flammability range. Thus, the effects of insulation on the risk of ignition of unburnt gases (smoke explosion) in an under ventilated environment is investigated. In this context, the experimental tests were carried out in a reduced scale cubic enclosure of 2m by 2m. The under-ventilated fire is modelled by a dodecane pool fire having a diameter of 40 cm and placed in the center of the enclosure's floor. An external ventilation system is mounted on the experimental device and provides an air supply rate with an air change per hour (ACPH) ranging from 3 to 5. The enclosure is depressurized by controlling an air intake and an air extraction ducts connected to a centrifugal fan. Two insulations are used to highlight the impact of the wall insulation on the risk of ignition of unburnt gases. The results show that a better insulated enclosure leads to faster fire growth, implying a greater peak in heat release rate, and therefore an increased risk of ignition. Thus, it is shown that the thermal insulation of the enclosure increases the rate of loss of mass of liquid fuel, but also reduces the rate of air supply to the intake duct due to the decrease in the level of vacuum in the compartment. As a result, the fire quickly becomes very under-ventilated and an accumulation of unburnt gases such as hydrocarbons, CO and H2 is observed. It is also shown that the ignition of unburnt volatiles at the inlet of the extraction duct occurs more easily when the compartment is more heat-tight with an ignition time of 16 min for the more insulated configuration for 21 min for the less insulated one.

Keywords: risk of ignition, ventilation system, insulation level, unburnt gases, smoke explosion

INTRODUCTION

Thermal incidents such as smoke explosions or backdraft have always attracted great interest in the field of fire safety engineering.

These result from explosions or gas deflagrations caused by re-ignition of unburnt gases in contact with fresh air supply during the extinction phase of the fire in the comaprtment. In order to improve the knowledge of this kind of phenomena, intensive research have been focused on the phenomenological analysis of backdraft in an enclosure after flame exhaust if there is a sudden supply of fresh air, e.g., due to a window or door opening or breaking, possibly due to the fire [1, 2]. The experimental study of Lassus et al. [3] showed that even in a highly confined setting, an extraction duct is an important pathway for the onset of re-ignition of a fuel rich mixture if there is a sudden supply of fresh air from an interconnecting ventilation system. Fundamental research [4-8] has been also carried on the phenomena of periodic oscillations and temporary flame quenching in a highly confined compartment equipped with mechanically ventilation system. The oscillation is caused by phenomenon of ghosting flame where the flame moves away from the pool and towards the air inlet [6]. The low-frequency oscillating flames with a drastic change in flame behavior were experimentally observed in a vitiated air enclosure [7, 8]. They postulated that issues involved for such oscillatory behavior are evaporation rate of condensed fuel [5], phenomena leading to a reverse flow in the supply ventilation system [4] and successive combustion of unburnt gases [5, 8].

As a condensed fuel, liquid pool fires usually generate a significant hazard in an enclosure due to large amount of unburnt volatile gases. The liquids could originate e.g., from leaks of transformers generators or other machinery. Knowing the liquid fire burning rate is the starting point of any fire safety analysis. Some empirical correlations [9-11] for mass loss rate of a liquid fuel is limited to the large enclosure where the heat fluxes from hot wall can be negligible. A global approach [12-14] for the mass burning flux of liquid fire is based on the measured radiative heat feedback to the condensed fuel surface or flame temperature. The burning rate of liquid pool can be significantly different in highly confined spaces, essentially attributed to the influence of air ventilation, heat radiation from hot walls and gas layer [15]. The existing empirical correlations [9-14] cannot take into account the factors such as oscillating flames and hot walls which affect significantly burning rate of the liquid pool fires in a highly confined enclosure. The quantitative agreement between experimental data and fully predictive simulation is again fair concerning the burning rate of liquid fuel in an enclosure [16]. Only using the mass loss rate collected experimentally, numerical simulation manages to correctly predict the thermal behaviour [17], visibility [18] and pressure [19] for the mechanically ventilated compartments.

In order to contain the potential release of radioactive material and avoid dispersion to the outside, nuclear installations are built with increasingly air/heat-tight envelopes [4]. Thermal insulation of enclosure is as such particularly relevant not only for Nuclear Power Plants [4], but also probably for an increasingly popular low-energy consumption type of building [20]. Heat-tightness in buildings envelopes has high potential in energy demand reduction and consequently in energy savings [21]. Air or heat-tight envelopes can sufficiently increase the peak overpressure and temperature from fast-growing fires to cause structural damage [19]. The increased pressure and temperature were identified [19, 21] as one of the consequences of better energy efficiency. The effect of heat permeability of enclosure has an importance on the fire behavior, notably on the thermal stratification of unburnt fuel gases [22]. The more heat-tight is a compartment, the more heat feedback is received to the surface of burning liquid fuel [23]. The re-ignition related risks may become more significant when the enclosure is more heat-tight due to a large formation of unburnt fuel. In this work, the investigation aims to identify experimentally the impact of thermal insulation of an enclosure on both the burning rate of liquid pool and ignition delay at entrance of the extraction duct.

EXPERIMENTAL SET-UP

The schematic experimental facility with a 2 m dimension in length/height and width is shown in Figure 1a, b, c. An external ventilation system consists of two intake ducts and an extraction duct. The intake was placed closer to the floor at a height of 0.3 m and the exhaust was placed near the ceiling at a height of 1.7 m. The extraction duct has a length of 1.1 m, and the dilution duct has a length of 2.5 m. Ducts have a square section of $0.2 \times 0.2\ m^2$.

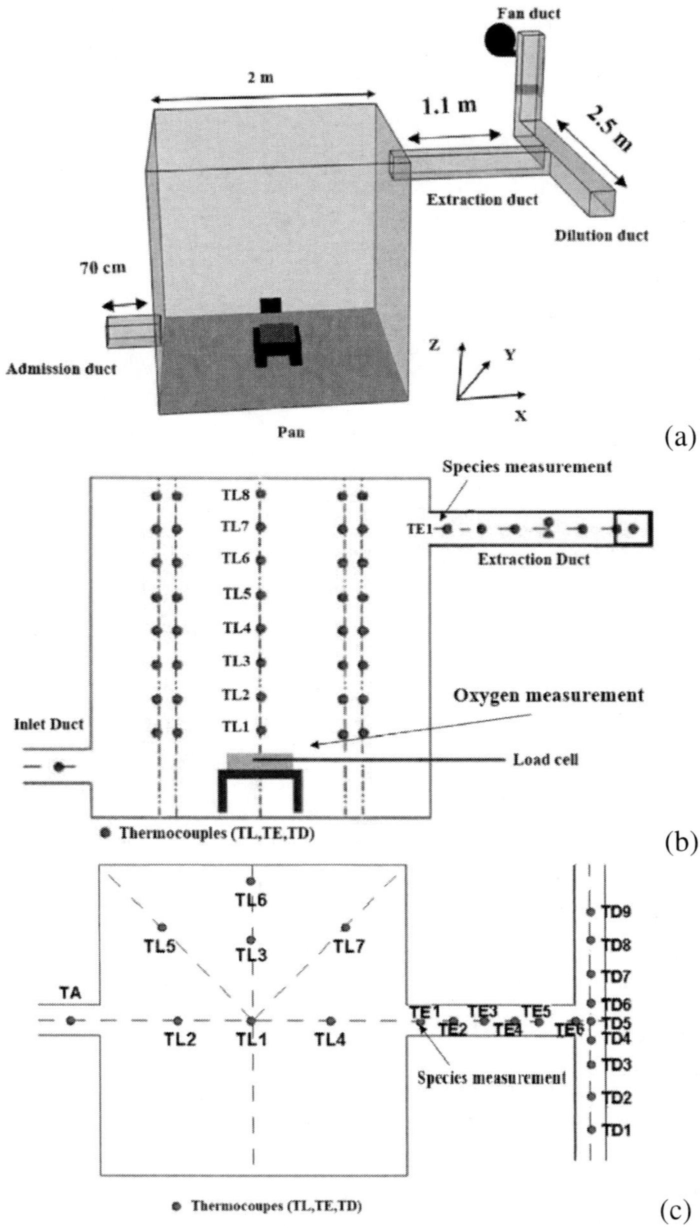

Figure 1. Mechanically ventilated enclosure fire connected to an extraction duct : (a) 3D view of the experimental facility; (b) side (x,z) view for the measurement locations and (c) measurment positions from the top (x,y) view.

Theoretically, the total flow loss in the duct from wall friction and duct fittings are assigned as dimensionless loss numbers to the ducts, as:

$$K = \frac{f \times L}{D} \tag{1}$$

where D and L are the duct diameter and length, respectively. The parameter, f, can be estimated from an approximation [24]. In the current work, all ventilation ducts are made of smooth galvanized steel, and can be considered as non-rough. This allows to neglect the regular pressure losses, and consider only the singular head losses. The air flow rates at the intake and dilution ducts are regulated as a function of the depression level by means of an adjustment of aeraulic resistances with rectangular flaps located upstream of these ducts. The experimental facility is not perfectly airtight mainly due to passage of the thermocouples through the wall and door gap. A leak of area in the order of 12–14 cm^2 [25] was estimated from conservation of the volume flow rate at the admission and extraction ducts as a function of the depression level. The pressure difference between the inside and outside is adjusted only in non-fire conditions.

A mechanical extraction is carried out by using System-Air Euro S7440, providing a maximum volume flow rate of 720 m^3/h, and pressure difference of $\Delta P_{max} = 186$ Pa. This is a centrifugal fan with single-phase variable speed to select the desired airflow rate [25]. As shown in Figure 1a, the fan is placed at the outlet of a duct orthogonal to the dilution duct at a height of 2.6 m. The fan characteristic curve (i.e., delivered flow rate as a function of pressure) is shown in Figure 2. Before activating the fire, a period of 10 min was included for establishing a depression below atmospheric pressure inside the enclosure. The depression level depends on the Air Change Per Hour (ACPH) of the enclosure with a range from 3 to 5, corresponding to a volume flow rate of 24 and 40 m3/h, respectively. The air inlet velocity in non-fire conditions varies from 0.1667 m/s (3 ACPH) to 0.2778 m/s (5 ACPH). After activating the fire, velocity, temperature and pressure inside the enclosure as well as the ventilation ducts depend on the fire dynamic.

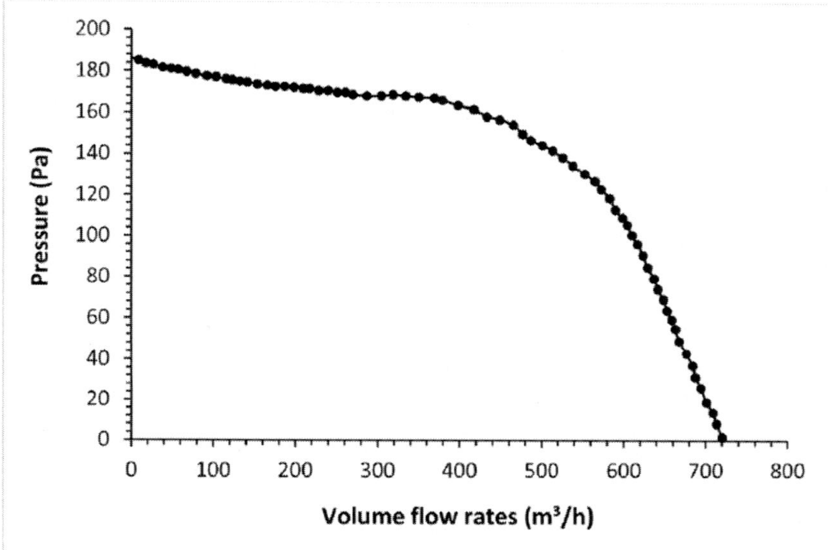

Figure 2. Relationship between the pressure and the free volume flow rate of the fan.

In previous work [25], a circular heptane or dodecane pan with a diameter varying from 23 to 40 cm was employed. A series of fire experiment were conducted by combining the three parameters such as pool size, ACPH and fuel type for investigating the extinction/ignition phenomena. In this work, a stainless-steel circular pan with a diameter of 40 cm was chosen. The liquid fuel pan was placed in the middle of the enclosure, slightly elevated at a height of 0.3 m. Dodecane is typically used as the fuel load, and occurrence of an auto-ignition was identified experimentally. The fuel mass was continuously monitored using a load cell, installed under the pan. The performed measurements include pressure, temperature and chemical species concentrations of gases filling the enclosure. A good repeatability was obtained with a difference below 2% regarding the mass loss rate during each fire tests. As shown in Figure 1b and c), the gas temperature was analyzed from different vertical profiles inside the enclosure, at the extraction duct and at the dilution one. Each of the vertical profiles comprises eight thermocouples positioned at regular intervals of 0.2 m over the height of the enclosure.

Two situations are taken into account for the intermediate levels of thermal insulation of enclosure. In the first situation, walls of the enclosure are made of 20-cm-thick cellular concrete with a thermal conductivity of $\lambda = 1$ W/m/K, a density of $\rho = 860$ kg/m3, and a specific heat of Cp = 0.88 kJ/kg/K. There is a relatively strong conductive heat loss via the walls during fire tests, and such case is named as 'heat loss'. In the second situation, the 20-cm-thick cellular concrete walls and ceiling are recovered by a 3.5-cm-thick insulating material as Promatech H with the following properties: $\lambda = 0.175$ W/m.K, $\rho = 400$ kg/m3 and Cp = 0.92 kJ/kg.K. In such case, the conductive heat loss via the walls becomes rather weak, and the enclosure is considered as 'heat tightness' or 'thermal insulation'. The reduction in the volume with an insulating material is low than 8%, and affects somewhat the amount of oxygen initially present in the enclosure. But the different wall thicknesses do not affect the oxygen supply rate by mechanical ventilation, and thus, the occurrence of auto-ignition of unburnt gas near the extraction duct.

Since experiments in full scale are expensive, the numbers of tests are limited. Fire compartment tests have been conducted at a reduced geometric scale into the PERICLES platform of the PPRIME Institute in France. Several scaling laws should be adapted to the studied problem. Reynolds number is large enough to ensure the turbulence of the flow. The Froude number, $Fr = u^2/gL$, is made constant by $u \propto L^{1/2}$ that comes from dimensionless variable of momentum conservation with a characteristic length L and velocity u. Flow time scale is obtained from mass conservation by $\tau \propto \frac{L}{u} \propto L^{1/2}$. Since the ventilation flow rate is the inverse of τ ,it scales with $L^{-1/2}$. From dimensionless variable of the energy equation, the source term contains the Zukoski number [26], defined as $\frac{\dot{Q}}{Fr \times L^{5/2} \times (\rho CpT) g^{1/2}}$.

In order to obtain a similitude between a full scale and a reduced scale compartments, a constant Froude number (Fr) must be maintained, and the ratio $\left(\dot{Q}^2/L^5\right)$ between heat release rate ($\dot{Q}$) and characteristic length (L) should be preserved. Based on this concept, our reduced scale experimental device with a volume of 8 m3 corresponds to a full scale 100 m3 compartment with a HRR below 1 MW.

RESULTS AND DISCUSSION

Histories of the relative pressure (P_0–P) for 3 and 5 ACPH are illustrated in Figure 3a, b. Here, P0 denotes the atmospheric pressure, and P the enclosure one. The fire-induced pressure can be said to follow three main stages: 1) activating the fire causes an overpressure inside the compartment; 2) with some time delay, the fully developed fire becomes ventilation controlled due to lack of oxygen with appearance of pressure oscillations; 3) occurrence of flame extinction in vitiated air enclosure makes a sudden increase of the depression level inside enclosure due to cooling effects. There is a tendency to enhance the fluctuations of the pressure with an increase of ACPH from 3 to 5. With thermal insulation, decreasing the thermal inertia $\left(\sqrt{\rho C p \lambda}\right)$ of the envelope by 70% led to a reduction in the depression peak by about 35%. During the fully developed fire, oscillations of the pressure become less sensitive to the intermediate levels of thermal insulation with an increase of ACPH to 5 (cf. Figure 3b). On the other hand, the strong oscillation of the pressure has been identified as a potential risk for the dynamic confinement in order to prevent from hazardous gas leaks.

The inlet velocity of fresh air supplied from the admission duct for 3 and 5 ACPH is presented in Figure 4(a, b). After activating the fire, an overpressure inside the compartment causes a reverse flow in the supply admission duct. During the fully developed fire, there is a tendency to enhance the fluctuations of the inlet flow velocity with an increase of ACPH from 3 to 5. At 3 ACPH, an overpressure appears with thermal insulation (cf. Figure 3a) around 750, leading to also a reverse flow in the supply admission duct with zero supply of fresh air. Finally, occurrence of flame extinction makes a strong supply of fresh air from the admission duct. With thermal insulation at 5 ACPH, the peak of inlet air velocity at the admission duct is sensitively reduced by a factor of 25%.

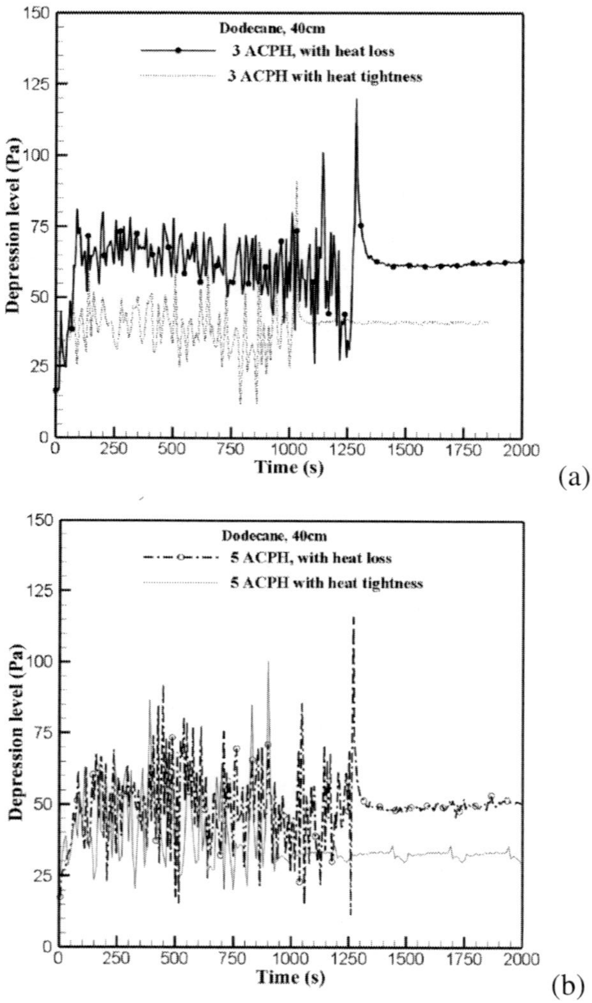

Figure 3. Impact of the intermediate levels of thermal insulation on the pressure inside enclosure : (a) at 3 ACPH and (b) at 5 ACPH.

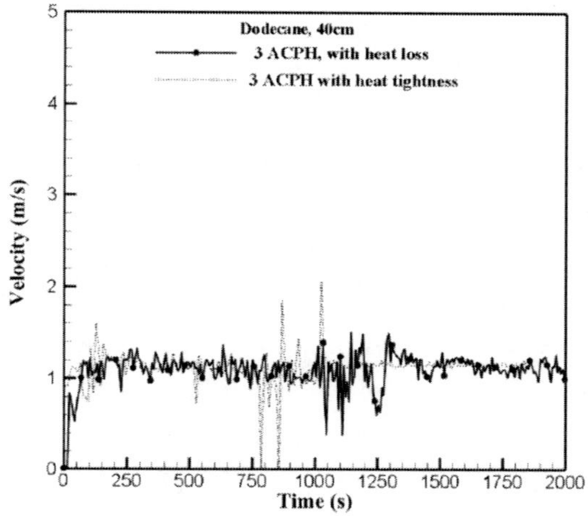

(a)

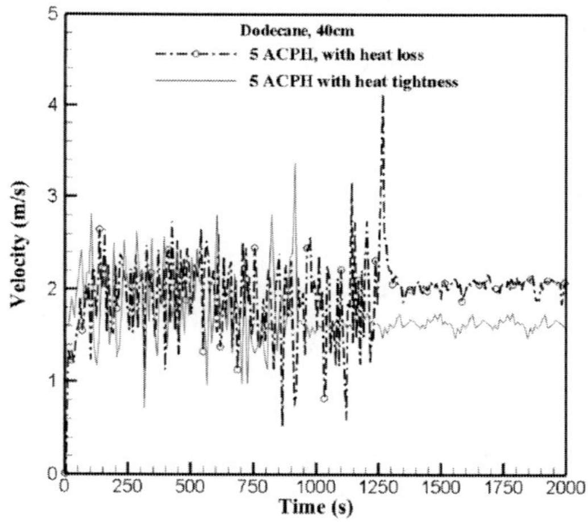

(b)

Figure 4. Impact of the intermediate levels of thermal insulation on the inlet air velocity at the admission duct : (a) at 3 ACPH and (b) at 5 ACPH.

Influence of the intermediate levels of thermal insulation of enclosure on the mass loss rate at 3 and 5 ACPH is illustrated in Figure 5. With heat loss, a rise of ACPH from 3 to 5 contributes to an increase of the liquid vaporization rate due to enhanced convective mass transfer over the liquid surface. As an illustration, the temperature profiles at TL4 (cf. Figure 1c) along the height at different instants are shown in Figure 6a, b, c. It can be seen that with thermal insulation of enclosure, the hot smoke layer (T > 200°C) descends rapidly to the level of liquid surface. As the upper smoke layer is established at 900 s, the hot gas temperature with heat tightness is about 2.5 times of that with heat loss. This results in a significant increase of the mass loss rate due to the enhanced radiative heat feedback from a hot gas layer. When the radiative heat transfer becomes dominant over the liquid surface, the liquid vaporization rate is practically insensitive to an increase of ACPH. In the initial growth phase, burning is fuel-controlled with a stable flame. This is followed by a fully developed, post-flashover phase with oscillations of the mass loss rate due to appearance of an unstable flame in a vitiated air enclosure [5, 15].

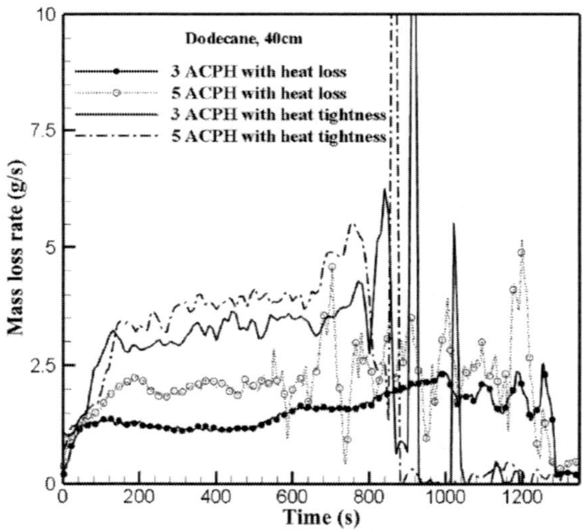

Figure 5. Impact of the intermediate levels of thermal insulation on the mass loss rate for different ACPH.

A typical evolution of the resulting heat release rate (HRR) at 3 and 5 ACPH is shown in Figure 7(a, b). Note that we lack the ability to accurately quantify the HRR by using oxygen calorimetry in such confined facility. The theoretical HRR is derived from the measured mass loss rate $\dot{m}_F$, as:

$$\dot{Q} = \dot{m} \times F \times \Delta Hc \tag{2}$$

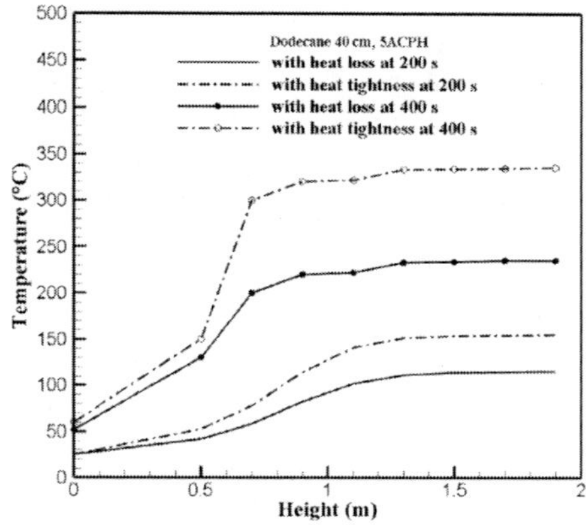

(a)

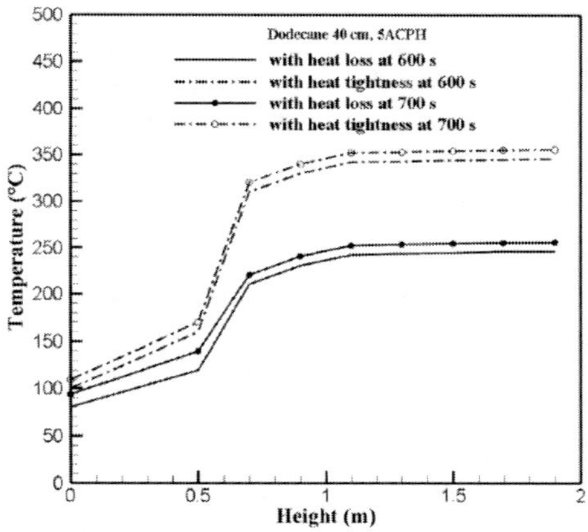

(b)

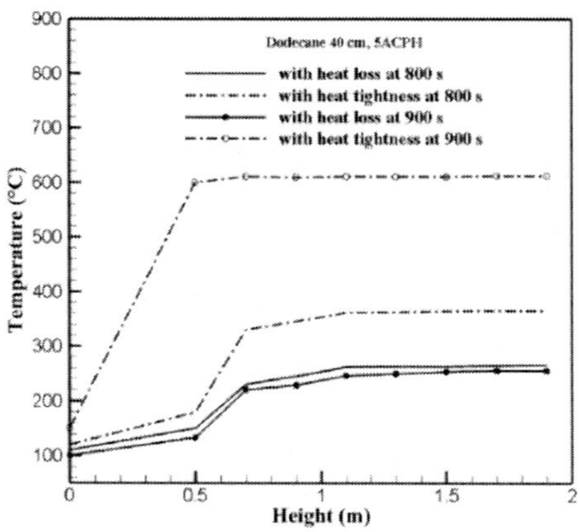

(c)

Figure 6. Impact of the intermediate levels of thermal insulation on the temperature profiles along the height at 5 ACPH : (a) at instants 200 s and 400 s; (b) at instants 600 s and 700 S and (c) at 800 s and 900 s.

Where, ΔHc denotes the energy released per kilogram of fuel consumed (44316 kJ/kg for dodecane). The effective HRR is calculated from the measured oxygen consumption rate at entrance of the extraction duct, as follows:

$$\dot{Q} = \dot{m}_A \left(Y_{O_2,\infty} - Y_{O_2} \right) \Delta H_O \tag{3}$$

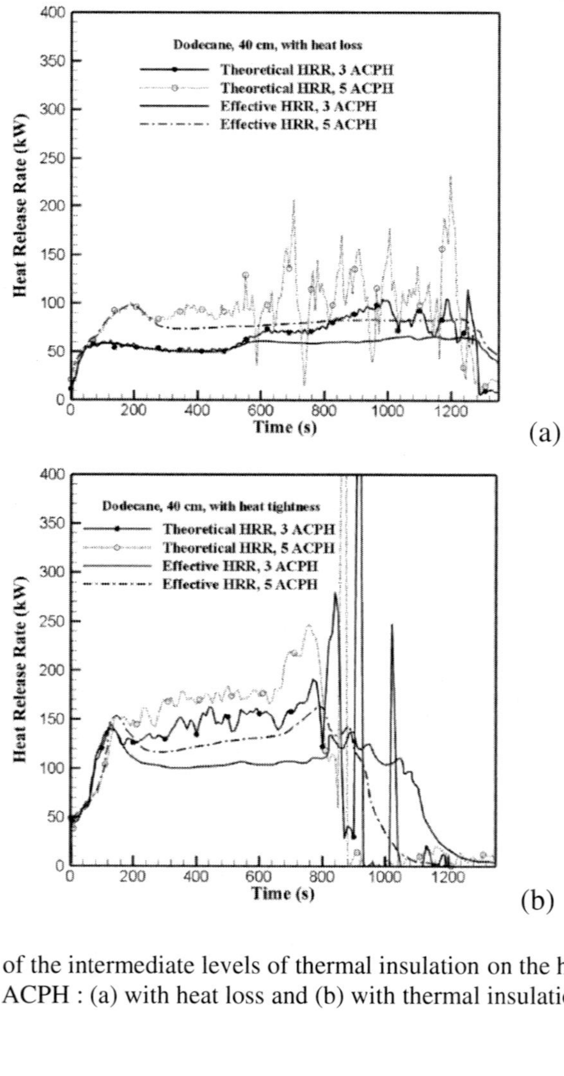

Figure 7. Impact of the intermediate levels of thermal insulation on the heat release rate for different ACPH : (a) with heat loss and (b) with thermal insulation.

Where, $\dot{m}_A$ denotes the inlet air flow rate (kg/s), Y_{O_2} the oxygen mass fraction and ΔH_O the energy released per kilogram of oxygen consumed (13100 kJ/kg).

Note that the boiling point of dodecane is about 220°C, and ignition of such liquid fuel needs a preheating process by using a torch. During preheating phase, heat flowing into the liquid interior over a pool size of 40 cm is uncontrollable, resulting in a difference in the early fire growth for the four fire scenarios. By assuming a quadratic growth for the four fire scenarios, an empirical formula is adopted [27] as follows:

$$\dot{Q} = a \times t^2 \tag{4}$$

where the early fire growth factor, a, is in kW/s2, and calculated from the growth time, t1, as . With heat loss, the growth time, t1, is about 160 s for 3 ACPH and 632 s for 5 ACPH, corresponding to a fast and slow fire growth, respectively. With thermal insulation, the growth time, t1, is practically identical at 3 and 5 ACPH with a value of about 320 s, corresponding to a medium fire growth. The oscillatory phenomena of the HRR are attributed to the coupling process between the fuel evaporation rate (cf. Figure 5) and the inlet air flow rate (cf. Figure 4), both attached to the intermediate levels of thermal insulation of enclosure. With heat loss of enclosure (cf. Figure 7a), the HRR curves show a gradual growing trend with a moderate peak of about 90 kW. The oxygen concentration initially present in the enclosure and the one supplied by mechanical ventilation entrain a sufficiently ventilated fire with an almost complete reaction. The effective HRR corresponds approximately to the theoretical one derived from the mass loss rate. It is observed that thermal insulation of enclosure leads to an ultra-fast fire growth (cf. Figure 7b), thus reducing the delay leading to flashover which is clearly identified by a rapid increase of HRR from 40 to 140 kW. The ultra-fast growing fire in a very consistent manner becomes progressively oxygen limited, that is to say ventilation-controlled, and the theoretical HRR is higher than the effective one. The extinction due to unstable combustion in a vitiated air enclosure happened earlier when the wall envelope is more heat-tight.

As an illustration at 5 ACPH, the histories of oxygen concentration at entrance of the extraction duct and near the floor in the vicinity of fire source are shown in Figure 8(a, b). It seems that during the initial growth phase, the oxygen concentration allows to entrain a sufficiently ventilated fire. For most fuels of interest for fire applications, the work of Huggett [28] indicates that combustion occurs when the local oxygen concentration is above a critical value, $Y_{O_2,\lim}$, evaluated from this expression:

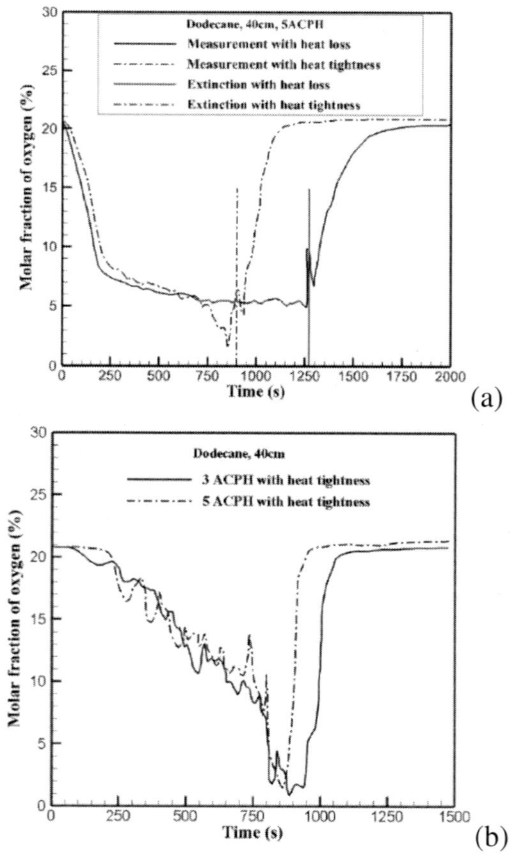

Figure 8. Impact of the intermediate levels of thermal insulation on the history of oxygen concentration : (a) at entrance of the extraction duct for 5 ACPH and (b) near the floor with thermal insulation for 3 and 5 ACPH.

$$Y_{O_2,\lim} = \frac{Cp\left(T_f - T_m\right)}{\Delta H_O} \tag{5}$$

The combustion products have an average specific heat Cp of 1.2 kJ/(kg.K), and the adiabatic flame temperature, Tf, is about 1600 K [29]. In the vicinity of fire source at a temperature of approximately Tm = 1000 K (cf. Figure 17), the limiting oxygen mass fraction is estimated to be YO2,lim = 0.05. This value is consistent with the visual identification of a visible flame extinction during each fire tests. The fire weakens earlier with thermal insulation and finally stops due to insufficient amount of oxygen even at 5 ACPH. After the fire extinction of the liquid fuel pan, the remaining flames as ghosting flame near the ceiling were not observed. Starting from 700 s, thermal insulation contributes to excessive combustion products, resulting in a decrease of oxygen concentration to about 2% in a fuel-rich upper layer near the extraction duct. Thermal insulation of enclosure induces a reduction in oxygen concentration by a factor of 60%. Occurrence of flame extinction makes a sudden increase of oxygen molar fraction to 7% at 900 s with heat tightness and 10% at 1250 s with heat loss due to a return of fresh air from dilution duct. This is followed later by a rise of the O2 concentration, mainly attributed to air supply thanks to mechanical ventilation. Evolution of the oxygen concentration near the floor (cf. Figure 8b) follows globally the trend near the extraction duct. Thanks to air supply from the admission duct, the oxygen concentration of about 10% is maintained near the floor till 750 s before reaching fire exhaust phase.

A relationship between the oxygen concentration near the floor and the one at entrance of the extraction duct can be established, as shown in Figure 9. This relationship presents three parts as a function of the fire regime. For a well-ventilated fire$\left(Y_{O_2} \geq 0.15\right)$, oxygen concentration near the floor and the extraction is almost the same. During the underventilated fire phase $\left(0.05 \leq Y_{O_2} \leq 0.15\right)$, a linear relation between the oxygen concentration near the floor and the one at the extraction duct is maintained. When the fire exhaust occurs $\left(Y_{O_2} \leq 0.05\right)$ (, an important slope for oxygen concentrations between the floor and the extraction duct is formed.

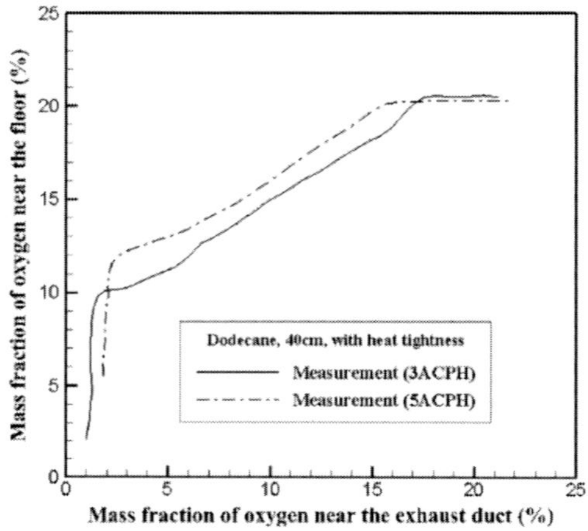

Figure 9. Relationship between the oxygen concentration near the floor and the one at entrance of the extraction duct.

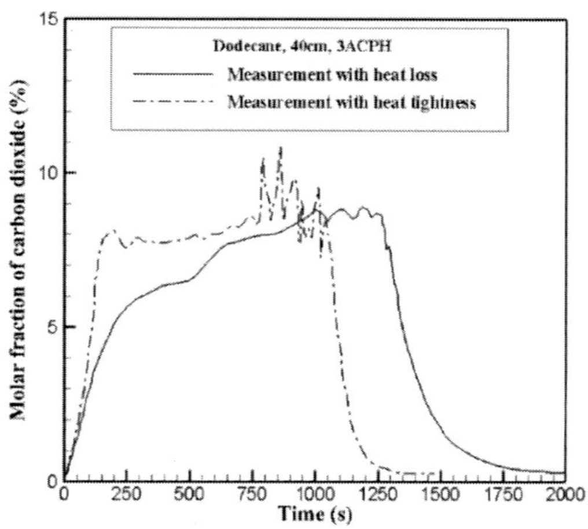

Figure 10. History of the molar fraction of carbon dioxide at the extraction duct for 3 ACPH.

Figure 10 shows the history of carbon dioxide concentration at entrance of the extraction duct for 3 ACPH. The CO_2 curve shows a smooth trend during the fully developed fire. When the wall envelope is very heat-tight, oscillations of carbon dioxide concentration are mainly attributed to occurrence of ignition near the extraction duct. Figure 11 shows a linear correlation of CO_2 as a function of oxygen molar fraction, formulated as:

$$[CO_2] = -0.65[O_2] + 0.13 \tag{6}$$

It seems that a molar fraction of oxygen above 10% is required to stabilize the fire behaviour with a linear increase of CO_2. At the stage where the fire becomes oxygen limited, the CO_2 growth rate decreases due to reduction in the chemical reactivity. Making the compartment more heat-tight does not change the trend of CO_2 formation as a function of O_2.

The empirical correlation for CO production is derived using the theoretical analyses in conjunction with experimental data [30, 31]. This correlation presents three fire regimes as a function of the oxygen concentration, as follows:

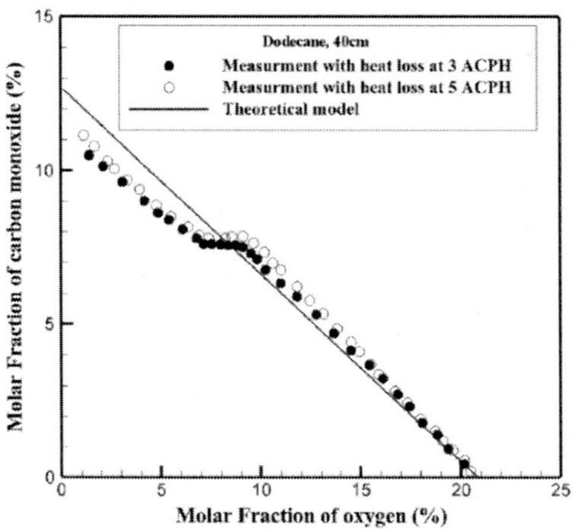

Figure 11. Evolution of carbon dioxide concentration versus oxygen concentration at the extraction duct for 3 and 5 ACPH with heat loss.

$$X_{CO} = \begin{cases} 0 & (XO2 \in [XMOC, XO2, \infty]) \\ -\frac{a}{XMOC} \cdot XO2 + a & (XO2 \in [XMOC,co, XMOC]) \\ -\frac{b}{XMOC} XO2 + b + a\left(1 - \frac{XMOC,co}{XMOC}\right) & \left(XO2 \in [0, XMOC,co]\right. \end{cases} \quad (7)$$

Where XMOC represents the Minimum Oxygen Concentration for a nearly complete reaction of a fuel-air mixture, and XMOC,co the Minimum Oxygen Concentration for oxidation of CO to CO2. The CO production is zero when the oxygen concentration is above XMOC, corresponding to a well-ventilated fire. The range of oxygen concentration between XMOC,co and XMOC implies an underventilated fire with a significant CO production due to an incomplete oxidation reaction. The oxygen concentration below XMOC,co results in a maximum CO production, implying a very under-ventilated fire. In Bryner's work [31], oxygen concentration can reach up to zero by using a gas burner to simulate the fire. In the current enclosure fire using a liquid fuel pan, the minimum oxygen concentration is about 2% with thermal insulation (cf. Figure 8).

The previous work [30] proposes also the following relationships as:

$$[CO] = c\sum[CmHn] \quad (8)$$

$$[CO] = [H_2] + d\sum C_m H_n \quad (9)$$

These correlations (8) and (9) allow to relate the hydrocarbons and hydrogen concentrations with the CO one. The empirical constants a, b, c and d are given in Table 1. It is found that such constants depend on the range $(0.7 < \varphi < 2)$ and $(2 < \varphi < 3)$ of the Global Equivalent Ratio (GER), defined as $\varphi = s \cdot \dot{m}_F / \dot{m}_A$, where s denotes the stoichiometric coefficient. The fuel supply rate $\dot{m}_F$, and air inflow rate $\dot{m}_A$ are controlled by the intermediate levels of thermal insulation via heat transfer and depression level in enclosure. The yields of CO and unburnt hydrocarbons, defined as the fraction of the fuel mass that is converted into these species, are commonly used as an indication for the level of underventilation in the enclosure [32]. To the best knowledge of the authors, only the yields of CO and unburnt hydrocarbons are a function of the GER in a well established upper layer of an enclosure fire [32]. In the current enclosure fire, there is unlikely to be sufficient time for two-layer establishment due to the strong fluctuations of HRR (cf. Figure 7). Moreover, identification of the ignition time delay controlled by the fire scenario depends on the time evolution of the local concentration of CO, H2 and unburnt hydrocarbons near the extraction duct.

Table 1. Empirical constants used in the correlations for CO, C_mH_n and H_2 concentrations

Parameter	With heat loss ($(0,7 < \varphi < 2)$	With heat tightness $(2 < \varphi < 3)$
X_{CMO}	11.1%	11.1%
$X_{CMO, CO}$	6.25%	6.25%
a	1.98	7.45
b	8	1.86
c	1.33	2
d	0.75	0.6

As illustrated in Figure 12(a, b), the empirically based correlation (7) for CO production is evaluated as a function of oxygen concentration at entrance of the extraction duct. Both the measurement and correlation indicate that thermal insulation enhances the CO production due to large vaporized fuel. As shown in Figure 13(a, b), the histories of carbon monoxide derived from correlation in transient mode are in a good agreement with the measured one within an error of 6% at 3 ACPH. However, a deviation of more than 40% from the measured CO concentration is found at 5 ACPH with heat loss.

This is probably attributed to an increasing oscillation of the depression level (cf. Figure 3b) under heat loss conditions, which enhances smoke fluctuations (turbulence effect) via leakage.

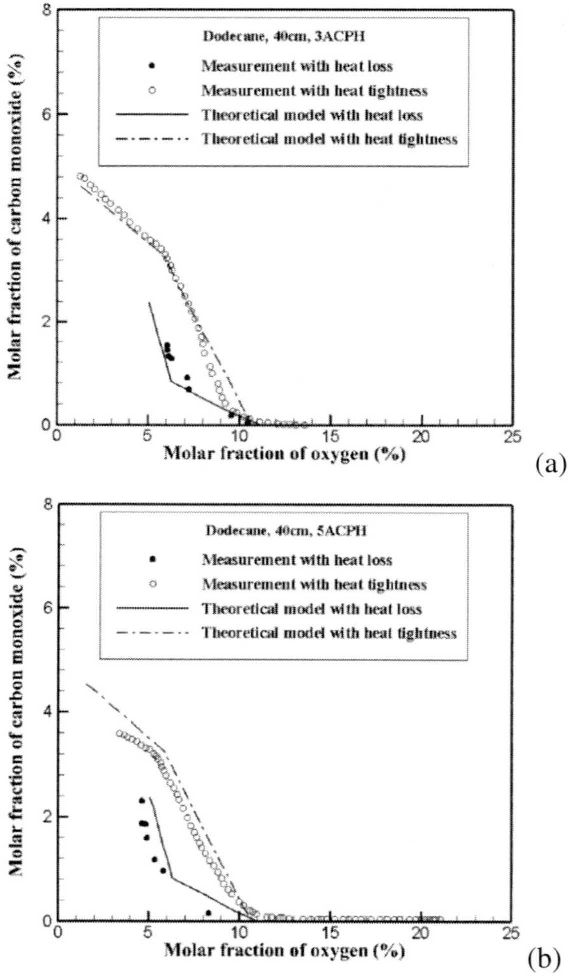

Figure 12. Evolution of carbon monoxide from measurement and correlation as a function of oxygen concentration: (a) at 3 ACPH and (b) at 5 ACPH.

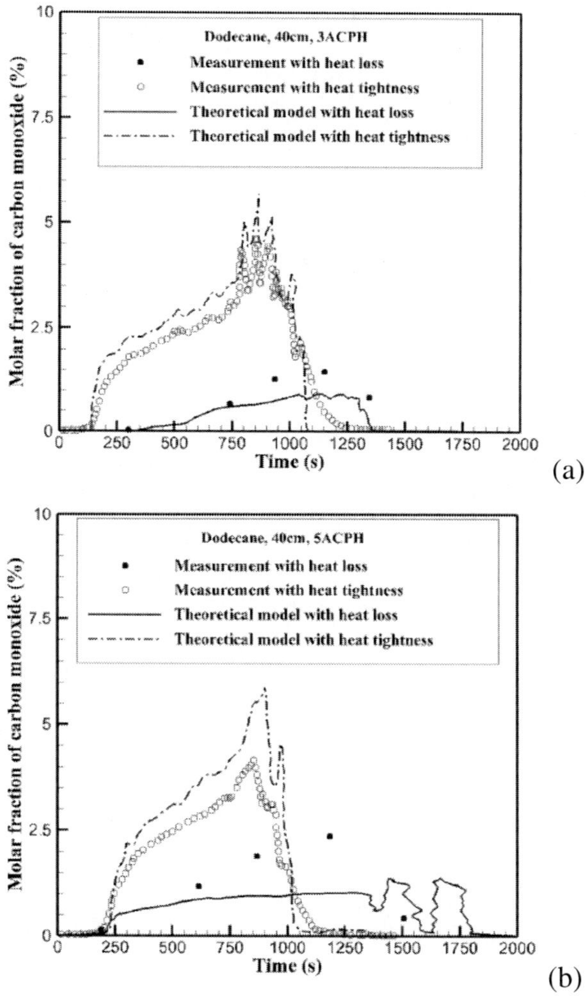

Figure 13. Effect of the intermediate levels of thermal insulation on carbon monoxide concentration in transient mode at entrance of the extraction duct : (a) at 3 ACPH and (b) at 5 ACPH.

Figure 14, Figure 15(a, b) show the histories of the chemical species such as hydrocarbons and hydrogen concentrations from the measurement and correlations (8) and (9) at entrance of the extraction duct.

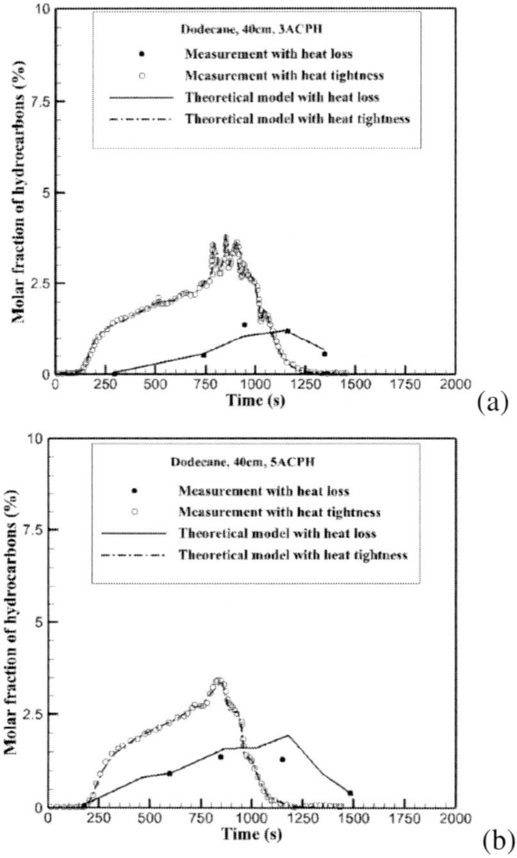

Figure 14. Effect of the intermediate levels of thermal insulation on hydrocarbons concentration in transient mode at entrance of the extraction duct : (a) at 3 ACPH and (b) at 5 ACPH.

The first period, corresponding to sufficiently ventilated conditions $(0 < \varphi < 0.7)$ [25] with low CO, H_2 and C_mH_n productions, lasts till 500 s with heat loss, and only 100 s with heat tightness of enclosure. During the second period, it is observed a gradual rise of unburnt fuel production with heat loss, and a rapid growth with heat-tightness. During the last period, corresponding to very under-ventilated conditions $(\varphi \geq 1.2)$ [25], thermal insulation contributes to the most severe production of unburnt fuels, even for 5 ACPH. The growth rate of the unburnt volatiles becomes faster when the enclosure envelope is very heat-tight, particularly during the post-flashover stage due to lack of oxygen. Globally, the attempt to use an empirical methodology has been successful for the calculation of the unburnt fuel production from the measured oxygen concentration.

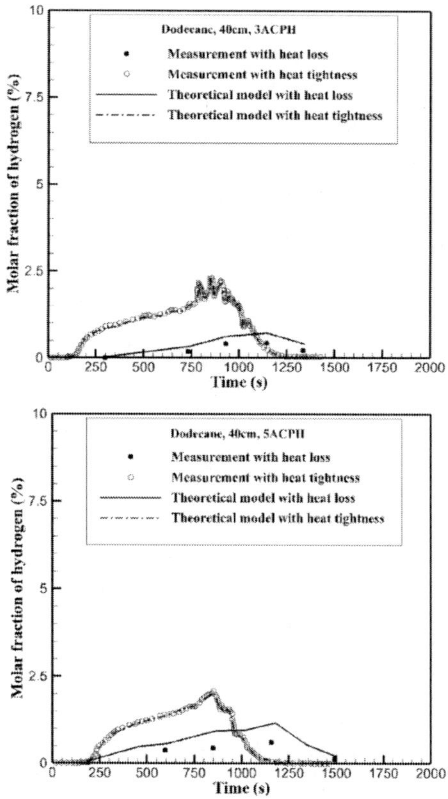

Figure 15. Effect of the intermediate levels of thermal insulation on hydrogen concentration in transient mode at entrance of the extraction duct: (a) at 3 ACPH and (b) at 5 ACPH.

The empirical correlations are predictive, and applicable for other fire scenarios [30], allowing to assess qualitatively the ignition risk of a fuel rich mixture near the extraction duct when a good similitude is preserved.

In the current enclosure, fire extinction of the liquid fuel pan occurs due to vitiated air with an oxygen concentration below 5% (cf. Figure 8). This section aims to verify theoretically and experimentally Babrauskas's extinction/ignition criteria [33], derived primarily from the Calorimeter apparatus. Theoretically, an auto-inflammation risk takes place when the unburnt fuel concentration is above the Low Flammability Limit (LFL) at the point of Auto-Ignition Temperature (AIT) [33]. The LFLi of each fuel and the LFL of gas mixtures as a function of both the fuel molar fraction Xi and the gas temperature are determined as follows:

$$LFL_i(T) = LFL_i(T_0)\left[1 - \frac{T - T_0}{1300 - T_0}\right] \text{ and}$$

$$LFL(T) = 100 \times \left[\sum_i \frac{X_i}{LFL_i(T)}\right]^{-1} \tag{10}$$

The Upper Flammability Limit (UFL) can be also calculated as:

$$UFL_i(T) = UFL_i(T_0)\left[1 - \frac{T - T_0}{1412 - T_0}\right] \text{ and}$$

$$UFL(T) = 100 \times \left[\sum_i \frac{X_i}{UFL_i(T)}\right]^{-1} \tag{11}$$

Table 2. Values of the AIT and LFL/UFL at T0 = 25°C for three fuel species

Parameter	CO	H2	Dodecane
LFL (T0 = 25°C)	12.5%	4%	0.6%
UFL (T0 = 25°C)	74%	75%	4.7%
AIT	588 °C	520 °C	208 °C

The values of the LFLi and UFLi at T0 = 25°C as well as AIT for three fuel species [33] are given in Table 2.

Note that even at a more developed stage with an enhanced evaporation rate, the concentration of the unburnt volatiles is never above the Upper Flammability Limit (UFL) with a value of about 20%. The molar fraction of unburnt volatiles in addition to the LFL of the gas mixtures near the extraction duct is presented in Figure 16a-d. According to the visual identification, occurrence of auto-ignition inside the extraction duct was observed experimentally at 3 or 5 ACPH with heat tightness (cf. Figure 16b, d) and only at 5 ACPH with heat loss (cf. Figure 16c). An auto-ignition of a fuel rich mixture with presence of combustion products near the extraction duct represents very complex chemistry and physical processes, there is still a lack of knowledge for complete understanding of such random phenomenon. In regards to ignition risk in a connected exhaust system, the three characteristic stages can be described, as follows:

i) In the early stages of a fire, starting from 350 s with heat loss and 200 s with thermal insulation, a stratified hotter unburnt fuels layer with a concentration above the Lower Flammability Limit (LFL) is formed near the ceiling. It seems that the point of Auto-Ignition Temperature of about 220°C for the primary fuel as dodecane (cf. Table 2) is not high enough for triggering an auto-ignition of a fuel rich mixture in addition to combustion products. It is noticeable that the auto-ignition temperature of a mixture is chemical composition-dependent (hydrocarbons, CO, H2, etc.), and difficult to be accurately quantified through bench-scale experiments.
ii) With a long time delay in a range of 16–21 min in the current study, the energy released per mass of oxygen consumed allows the smoke temperature to raise above 350°C, high enough for an auto-ignition of a mixture of the unburnt gases and vitiated air. However, during this stage, the large vaporized fuel due to enhanced heat transfer does not completely contribute to the heat release. Thus, a part of the fuel vapor accumulates with a concentration considerably higher than LFL, particularly with thermal insulation. Consequently, there is insufficient oxygen available near the extraction duct for triggering the ignition.

iii) Occurrence of flame extinction with an oxygen concentration below 5% makes a sudden increase of the depression level inside enclosure due to cooling effects. This induces a sudden supply of fresh air from dilution duct, characterized by a sharp increase of oxygen concentration to 10% (cf. Figure 8a) at the extinction point of about 1250 s with heat loss and to 7% at 900 s with heat tightness. The supply of fresh air leads to a rapid dilution of unburnt fuel with a concentration close to a stoichiometric fuel-air mixture. This is ideal for triggering the ignition of a fuel rich mixture with a peak reaching a typical flame temperature higher than 550°C. With heat loss of enclosure (cf. Figure 16c), occurrence of auto-ignition near the extraction duct was consistently observed at 5 ACPH with a longer time delay of about 1320 s after the extinction at 1250 s. Thermal insulation of enclosure (cf. Figure 16d) contributes to a significant reduction in the ignition delay time to about 780 s due to fast-growing fires.

It is found that the delay for the transition from extinction of the fuel pan to ignition near the extraction duct is about 60–70 s. With heat loss of the enclosure at 3 ACPH (cf. Figure 16a), after extinction at 1300 s, dilution of the unburnt fuel from the return of fresh air makes its concentration decrease below LFL during the transition delay. The temperature peak was not experimentally detected by the thermocouple near the extraction duct. This implies that an auto-ignition of the unburnt gases does not take place there, probably attributed to cooling of the smoke. Note that the response time of the thermocouples used in the experiment is close to a second by including thermal mass inertia and accumulation of soot particles.

Besides, raw thermocouple data are corrected by heat transfer effects via convection and radiation. During auto-ignition of the unburnt gases, accurate measurement of flame temperature of an auto-ignition in random manner by using thermocouples is difficult by taking into account a multitude of potential errors, including surface reactions, radiation, convective gain or loss. The leakage of the experimental facility is somewhat uncontrollable during each fire tests.

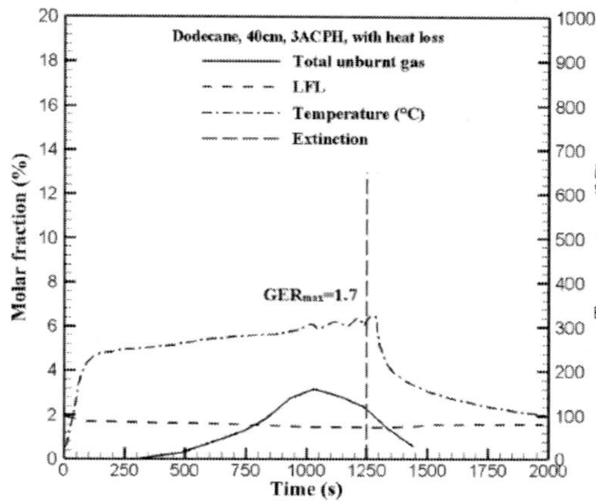

(a)

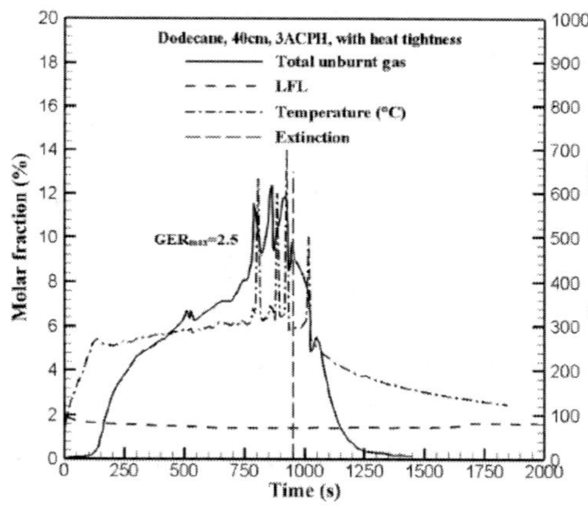

(b)

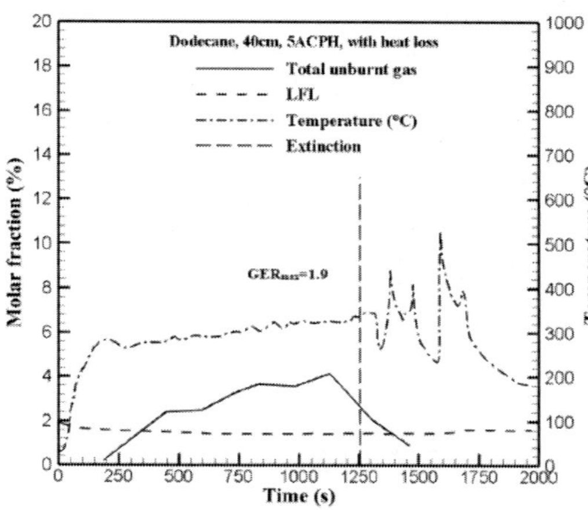

(c)

Figure 16. (Continued)

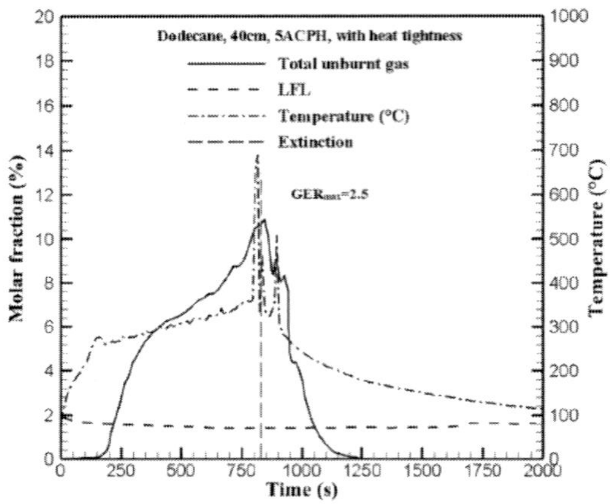

(d)

Figure 16. Risk assessment of unburnt gas ignition at entrance of the extraction duct at 3 ACPH and 5 ACPH: (a)-(c) with heat loss and (b)-(d) with heat tightness.

Globally, the compartment fire dynamics depend on the Global Equivalence Ratio (GER) associated with pool size, ACPH and fuel type. In a specific configuration with an admission duct placed closer to the floor and an exhaust near the ceiling, the GER above 1.5 represents a set of dangerous conditions for spontaneous ignition at entrance of the extraction duct (cf. Figure 16a–d). The auto-ignition of unburnt fuels at the extraction duct occurs without needing a pilot flame as ghosting flame. To the knowledge of the authors, any attempt to draw a general conclusion for all configurations is discouraged. In order to obtain a good similitude between various scale compartments, a preservation of the ratio between heat release rate and characteristic length with a constant Froude number is necessary.

Figure 17 (a) and (b) shows the histories of gas-phase temperature at a thermocouple tree in the center of the compartment at a height of 60 cm (TL3, cf. Figure 1b) from the liquid surface. It is seen that thermal insulation of the wall envelope at 3 ACPH causes a strong decreasing trend of the temperature from 120 s. This is attributed to the flame moving from the fuel source due to lack of oxygen, expanding radially and horizontally towards the air inlet [6]. An inverse dependence of the gas temperature with time is observed.

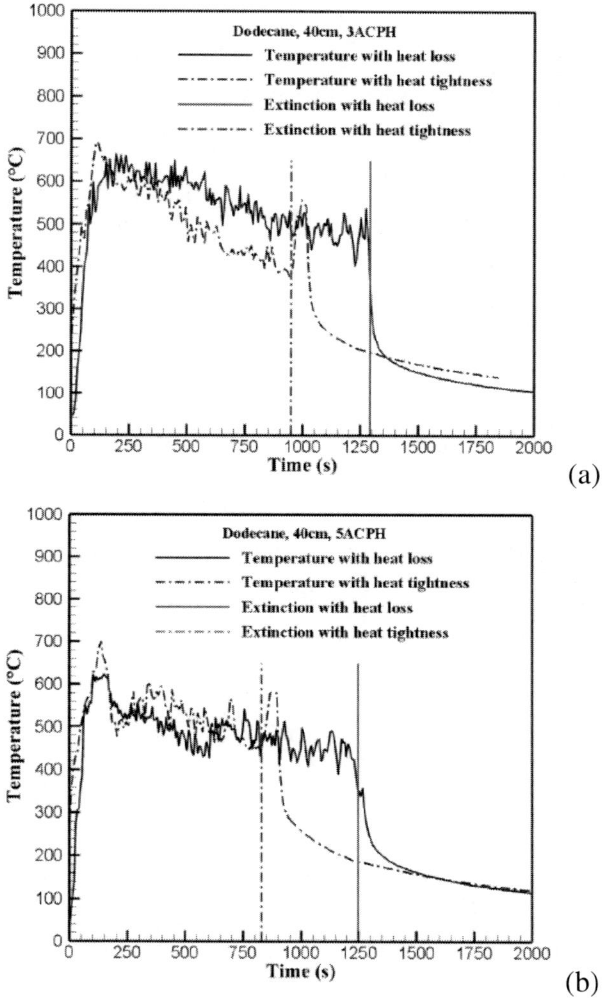

Figure 17. Influence of heat permeability on the history of gas-phase temperature in the center of the compartment above the liquid pool fire: (a) at 3ACPH and (b) at 5 ACPH.

This is contrary to the trend at the extraction duct where the gas temperature rises with time due to energy accumulation. The auto-ignition near the extraction duct at 900 s (cf. Figure 17a) makes an oscillation of the smoke layer temperature, and its impact on the gas temperature in the vicinity of fuel pan is characterized by a sharp temperature spike. An increase of ACPH to 5 (cf. Figure 17b) helps maintain a relatively high temperature level for many minutes, characterized by a slightly decreasing trend. As shown in Figure 6c, thickness of the established hot smoke layer ($T > 200°C$) decreases from 1.9 m with thermal insulation to 1.3 m with heat loss. Thus, impact of the temperature oscillation within the smoke layer due to ignition near the extraction duct on the centerline gas temperature seems less remarkable with heat loss. Finally, a sharp decreasing trend of the gas temperature from 20 min with heat loss and 16 min with heat tightness corresponds to the onset of extinction.

CONCLUSION

In this chapter, it is shown that the insulation of the enclosure can highly increase the liquid vaporization rate, thus reducing the air supply flow rate due to a decrease of depression level inside of the enclosure. The experimental tests were carried out in an enclosure connected to a ventilation network allowing to have a depression inside of the compartment. From the phenomena of expansion of the fire, the depression decrease induces a reduction of the air flow. Moreover, after consuming the initial amount of air inside of the enclosure, the combustion process is quickly ventilation controlled, causing the under-ventilated fire to progress to a big fire such as the one studied in this chapter. For these conditions, a significant amount of unburnt volatiles is producted and accumulate in the extraction duct. If the concentration of unburnt volatiles is in the flammability range (higher than LFL), and the fire has been established for sufficiently long (in this study, about 15 min), there is a risk of ignition inside of the extraction duct. In addtition, when the concentration of unburnt gases is outside of the flammability range (higher than UFL), the ignition of the hotter unburnt fuels at the entrance of the extraction duct is attributed to a return of air from dilution duct at the stage of fire extinction, which induces a sudden increase of depression level inside of the enclosure. An ignition can happen early near the extraction duct, as long as the enclosure remains sufficiently insulated. Globally, the causes involved for the occurrence of auto-ignition near the extraction duct are: amount of accumulated unburnt volatiles above LFL, gas temperature above 350°C, oxygen concentration above 7% with high insulation and 10% with low insulation, and fire extinction with an oxygen concentration below 5%.

To conclude, the levels of thermal insulation of an enclosure appears to be a critical factor for the delay of occurrence of ignition, along with the vaporization rate of condensed fuel and the depression level inside enclosure. The work is continuing with the aim of improving the evaluation of auto-ignition temperature near the extraction duct by analyzing a more detailed chemical species. The future investigation also aims to identify the scale effect of an enclosure on heat fluxes over liquid surfaces via the radiation contribution from the flame, the smoke layer and the hot walls.

ACKNOWLEDGMENTS

The authors would like to thank the AREVA industry partner for its involvement in the project. This work was supported by the CPER FEDER project of Région Nouvelle Aquitaine, France.

REFERENCES

[1] Gojkovic, D. (2000). *Initial backdraft experiments.* Report 3121. Department of Fire Safety Engineering, Lund University, Sweden.

[2] Horvat, A., & Sinai, Y. (2007). Numerical simulation of backdraft phenomena. *Fire safety journal*, 42(3), 200-209.

[3] Lassus, J., Studer, E., Garo, J. P., Vantelon, J. P., Jourda, P., & Aine, P. (2010). Influence of ventilation on ignition risk of unburnt gases in the extraction duct of underventilated compartment fire. *Combustion science and technology*, 182(4-6), 517-528.

[4] Prétrel, H., Le Saux, W., & Audouin, L. (2012). Pressure variations induced by a pool fire in a well-confined and force-ventilated compartment. *Fire safety journal*, 52, 11-24.

[5] Zavaleta, P., & Audouin, L. (2018). Cable tray fire tests in a confined and mechanically ventilated facility. *Fire and Materials*, 42(1), 28-43.

[6] Wahlqvist, J., & Van Hees, P. (2013). Validation of FDS for large-scale well-confined mechanically ventilated fire scenarios with emphasis on predicting ventilation system behavior. *Fire Safety Journal*, 62, 102-114.

[7] Hu, Z., Utiskul, Y., Quintiere, J. G., & Trouve, A. (2005). A comparison between observed and simulated flame structures in poorly ventilated compartment fires. *Fire Safety Science*, 8, 1193-1204.

[8] Takeda, H. (1985). Oscillatory phenomenon and inverse temperature profile appearing in compartment fires. *Combustion and flame*, 61(1), 103-105.

[9] Peatross, M. J., & Beyler, C. L. (1997). Ventilation effects on compartment fire characterization. *Fire Safety Science*, 5, 403-414.

[10] Utiskul, Y. (2006). *Theoretical and experimental study on fully-developed compartment fires* (Doctoral dissertation).

[11] Tewarson, A., Lee, J. L., & Pion, R. F. (1981, January). The influence of oxygen concentration on fuel parameters for fire modeling. In *Symposium (international) on combustion* (Vol. 18, No. 1, pp. 563-570). Elsevier.

[12] Orloff, L., & De Ris, J. (1982, January). Froude modeling of pool fires. In *Symposium (International) on Combustion* (Vol. 19, No. 1, pp. 885-895). Elsevier.

[13] Hamins, A. P., Yang, J. C., & Kashiwagi, T. (1999). *Global model for predicting the burning rates of liquid pool fires* (NISTIR 6381) (No. NIST Interagency/Internal Report (NISTIR)-6381).

[14] Beaulieu, P. A., & Dembsey, N. A. (2008). Effect of oxygen on flame heat flux in horizontal and vertical orientations. *Fire Safety Journal,* 43(6), 410-428.

[15] Vilfayeau, S., Ren, N., Wang, Y., & Trouvé, A. (2015). Numerical simulation of under-ventilated liquid-fueled compartment fires with flame extinction and thermally-driven fuel evaporation. *Proceedings of the Combustion Institute*, 35(3), 2563-2571.

[16] Sikanen, T., & Hostikka, S. (2017). Predicting the heat release rates of liquid pool fires in mechanically ventilated compartments. *Fire Safety Journal*, 91, 266-275.

[17] Merci, B., & Van Maele, K. (2008). Numerical simulations of full-scale enclosure fires in a small compartment with natural roof ventilation. *Fire safety journal*, 43(7), 495-511.

[18] Węgrzyński, W., & Vigne, G. (2017). Experimental and numerical evaluation of the influence of the soot yield on the visibility in smoke in CFD analysis. *Fire Safety Journal*, 91, 389-398.

[19] Hostikka, S., Janardhan, R. K., Riaz, U., & Sikanen, T. (2017). Fire-induced pressure and smoke spreading in mechanically ventilated buildings with air-tight envelopes. *Fire Safety Journal*, 91, 380-388.

[20] Cabeza, L. F., Castell, A., Medrano, M., Martorell, I., Pérez, G., & Fernández, I. (2010). Experimental study on the performance of insulation materials in Mediterranean construction. *Energy and Buildings*, 42(5), 630-636.

[21] Fourneau, C., Cornil, N., Delvosalle, C., Breulet, H., Desmet, S., & Brohez, S. (2012). Comparison of fire hazards in passive and conventional houses. *Chemical Engineering Transactions*.

[22] Sinai, Y. L. (1999). Comments on the role of leakages in field modelling of under-ventilated compartment fires. *Fire safety journal*, 33(1), 11-20.

[23] Nasr, A., Suard, S., El-Rabii, H., Garo, J. P., Gay, L., & Rigollet, L. (2013). Heat feedback to the fuel surface of a pool fire in an enclosure. *Fire safety journal*, 60, 56-63.

[24] Zigrang, D. J., & Sylvester, N. D. (1985). *A review of explicit friction factor equations,* 280-283.

[25] Magnognou, B., Garo, J. P., Coudour, B., & Wang, H. Y. (2017). Risk analysis of unburnt gas ignition in an exhaust system connected to a confined and mechanically ventilated enclosure fire. *Fire Safety Journal*, 91, 291-302.

[26] Zukoski, E. E. (1986). Fluid dynamic aspects of room fires. In *Fire Safety Science, Proceedings of the First International Symposium* (pp. 1-30).

[27] *NFPA (National Fire Protection Association) Standard Growth Curves,* BRE-Project (2009).

[28] Huggett, C. (1980). Estimation of rate of heat release by means of oxygen consumption measurements. *Fire and Materials*, 4(2), 61-65.

[29] Beyler, C. (2016). Flammability limits of premixed and diffusion flames. *In SFPE handbook of fire protection engineering*, 529-553. Springer, New York, NY.

[30] Lassus, J., Courty, L., Garo, J. P., Studer, E., Jourda, P., & Aine, P. (2016). Estimation of species concentrations during a fire in a reduced-scale room. *Journal of Fire Sciences*, 34(1), 30-50.

[31] Bryner, N. P., Johnsson, E. L., & Pitts, W. M. (1994). Carbon Monoxide Production in Compartment Fires: Reduced-Scale Enclosure Test Facility, *NIST Interagency/Internal Report* (NISTIR-5568).

[32] Hurley, M. J., & Rosenbaum, E. R. (2016). Performance-based design. In *SFPE handbook of fire protection engineering*, 1233-1261. Springer, New York, NY.

[33] Babrauskas, V. (2003). Ignition Handbook: Principles and Applications to Fire Safety Engineering, Fire Investigation. Risk Management and Forensic Science. *Fire Science Publishers*, 41-140.

BIOGRAPHICAL SKETCH

Brady Manescau, PhD

Affiliation: INSA Centre Val de Loire (National Institute of Applied Sciences), University of Orléans, PRISME EA 4229 Laboratory, France

Education: PhD., M.Eng.Sci., B.E.

Business Address: 88 boulevard Lahitolle CS 60013 18022 Bourges, France

Research and Professional Experience:

Brady Manescau is associate Professor at National Institute of Applied Sciences and Researcher in fire sciences and fire safety in the Combustion and Explosion department of the PRISME EA 4229 Laboratory.

Publications from the Last 3 Years:

1. Manescau B., Courty L., Acherar L., Coudour, B., Wang, H.Y., Garo, J.P., 2020. Effects of ventilation conditions and procedures during a fire in a reduced-scale room. *Process Saf. Environ. Prot.* 144, 263–272. https://doi.org/10.1016/j.psep.2020.07.035
2. Manescau B., Chetehouna, K., Sellami, I., Nait-Said, R., & Zidani, F. (2020). BLEVE Fireball Effects in a Gas Industry: A Numerical Modeling Applied to the Case of an Algeria Gas Industry. *Fire Safety and Protection*. IntechOpen.
3. Manescau B., Chetehouna, K., Serra, Q., Wang, A., & Florentin, E. (2020). Numerical Study on the Outdoor Wind Effects on Movement Smoke along a Corridor. In *Fire Safety and Protection*. IntechOpen.
4. Manescau B., Wang, H. Y., Coudour, B., & Garo, J. P. (2019). Influence of heat tightness of an enclosure fire on ignition risk of unburnt gases in a connected exhaust system–An experimental study. *Fire Safety Journal*, 109, 102867.
5. Grange, N., Manescau, B., Chetehouna, K., Gascoin, N., Lamoot, L., Coppalle, A., ... & Reynaud, I. (2019). Fire resistance of carbon-based composite materials under both ideal and realistic normative configurations. *Applied Thermal Engineering*, 159, 113834.
6. Wang, A. J., Manescau B., Serra, Q., Chetehouna, K., Florentin, E., & De Izarra, C. (2019). Numerical simulations of outdoor wind effects on smoke spreading along a corridor: Physical and sensitivity analysis. *International Journal of Thermal Sciences*, 142, 332-347.
7. Dong, H. Q., Magnognou B., Garo, J. P., Coudour, B., & Wang, H. Y. (2019). Fully Coupled CFD Simulation of the Ignition Risk of

Unburnt Gases in an Exhaust System Connected to a Mechanically Ventilated Enclosure Fire. *Fire Technology,* 55(1), 149-174.

8. Sellami, I., Manescau, B., Chetehouna, K., de Izarra, C., Nait-Said, R., & Zidani, F. (2018). BLEVE fireball modeling using Fire Dynamics Simulator (FDS) in an Algerian gas industry. *Journal of Loss Prevention in the Process Industries*, 54, 69-84.
9. Gascoin, N., Manescau, B., Akridiss, S., & Chetehouna, K. (2018). Solubility of nitrogen into jet fuel. *Industrial & Engineering Chemistry Research*, 57(7), 2441-2448.
10. Magnognou (Manescau), B., Garo, J. P., Coudour, B., & Wang, H. Y. (2017). Risk analysis of unburnt gas ignition in an exhaust system connected to a confined and mechanically ventilated enclosure fire. *Fire Safety Journal*, 91, 291-302.
11. H. Q. Dong, J. P. Garo, B. Magnognou (Manescau), H. Y. Wang, A tractable solution for engineering calculations On Buoyancy Dominated Turbulent non-premixed Flames, *Fire Science and Technology*, Vol. 34, pp. 1-21, 2015.
12. H. Y. Wang, H. Q. Dong; B. Magnognou (Manescau); J. P. Garo; B. Coudour, Fully coupled CFD simulation of the ignition risk of unburnt gases in an exhaust system connected to a mechanically ventilated enclosure fire, *Fire Technology*, 2018.

In: Critical Issues and Analysis … ISBN: 978-1-53618-738-0
Editors: A. Cablé and B. Manescau

Chapter 2

VISIBILITY THROUGH SMOKE AND ITS EVALUATION FOR FIRE SAFETY ENGINEERING DESIGN

Yaping He[*]
School of Computing, Engineering and Mathematics
Western Sydney University, Sydney, Australia

ABSTRACT

Visibility is a crucial factor affecting human behaviour and safety during fire emergency evacuations. Adequate evaluation of visibility through smoke will assist in the design of fire safety systems, particularly the smoke control and signage systems, in built environment to maintain tenable conditions and guide occupants to safety during fire emergencies. In this chapter, an appraisal of previous studies on visibility through smoke is presented. Some fundamental concepts and physics of light transmission are discussed. Based on the review, a theoretical framework is established for attenuation of light from two kinds of sources that resemble the warning and exits signs commonly and mandatorily installed in built environments. This framework incorporates both the divergence effect of light rays from light-emitting sources and attenuation effect by smoke. Light sources for emergency warning and signage are categorised into point sources and

[*] Corresponding Author's Email: y.he@westernsydney.edu.au.

panel sources. For a point source the simple differential equation for divergence effect is merged with that accounting for the attenuation effect by smoke. The straightforward solution for light intensity is presented as a function of distance from the source as well as smoke extinction coefficient. For a light emitting panel, the concept of view factor in the radiation heat transfer theory is applied to account for variation in light intensity (luminance) as a function of position (distance and angle) relative to the panel. The resultant function is moderated by the smoke attenuation factor to give the total variation in luminance. A method is then developed for the evaluation of visibility of warning lights and exit signs in heterogeneous smoky conditions. This method is based on an empirical correlation obtained by previous researchers in the literature and with further consideration of attenuation of luminance along the line of sight from the observer to the light source. The application of this method can be extended to the processing of computational fluid dynamics simulation results to obtain visibility of exit signs through smoke. The advantage and limitations of the new method are discussed. Recommendations for future research are presented at the end.

Keywords: attenuation, divergence, extinction coefficient, heterogeneity, optical density, soot

NOMENCLATURE

A	area (m^2)
C	constant
D	attenuation coefficient (m^{-1})
E	visibility factor
F	view factor
I	light intensity, irradiance (Wm^{-2})
I_o	incident light intensity (Wm^{-2})
k	extinction coefficient (m^{-1})
L	luminance (cdm^{-2})
l	distance (m)
O_p	opacity
P	emitting power (W)
r	radial distance (m)
S	visibility, distance (m)

Greek

β angle

Subscript

b background
o object

Introduction

Visibility is a crucial factor in determining tenability of the conditions in built environment involved in fires and in evaluating the required safe evacuation time for building occupants in case of fire or other emergencies [1]. The smoke generated by fires can obscure the vision of building occupants. Reduction in visibility could cause reduction in their walking speed and prolong the evacuation process [2-4]. Further reduction or loss of visibility combined with other effects of smoke may lead to disorientation and entrapment of building occupants [5] and even cause fatality. A study carried out by Bryan [6] showed that visibility of exit signs, doors and windows is of great importance to building occupants attempting to evacuate a building. The objectives of building fire safety design and assessment are concerned with maintaining a tenable condition with adequate visibility to allow safe evacuation by building occupants in the event of a fire, and to facilitate fire brigade intervention [7]. Visibility of signage is also a subject of study of safety in conditions other than the fire affected environment [8].

The modern day performance based fire safety engineering and fire safety science take advantage of computer aided design and analysis, and make use of computational fluid dynamics simulations to estimate the consequences of fire incidences in order to assess the effectiveness of a proposed fire safety engineering design [9], investigate particular flow phenomenon [10] or conduct fire forensic studies [11, 12]. Computational fluid dynamics fire models predict thermodynamic, chemical and fluid flow properties, such as temperature, pressure, velocity, species concentrations and soot aerosol density by solving the governing equations. Some software programs convert these properties into the parameters, such as visibility, that can be used for engineering design and assessment purpose. This conversion process is an important step to practical use of the models.

The study of visibility of exits and exit signs has attracted the attentions of fire safety scientists since the second half of the last century. Generally, the study falls into two areas. The first focuses on the relationship between fires and visibility and the second on the effect of visibility on human behaviour [4, 13, 14]. The focus of the current chapter is in the first area. More specifically, how to translate smoke properties and their distribution in spaces into the visibilities of warning lights and signage through smoke is the primary concern of the current chapter.

Much research has been conducted over the years, which resulted in valuable information on visibility through smoke. Empirical correlations were obtained between visibility and smoke properties [1, 2]. The correlations developed by Jin [2] have been adopted in computer models [9] and widely used in engineering applications.

Visibility is an extensive parameter which depends on the dimensions of the domain or the conditions along the line of sight between a subject (building occupant) and an object (exit sign). Jin's equation correlates visibility with smoke (light) extinction coefficient which is an intensive parameter. This correlation is valid only under the circumstance of uniform or homogeneous distribution of the extinction coefficient, but may not give appropriate estimates of visibility in heterogeneous smoke distribution environments which are more likely to occur in real fire scenarios. The issue of visibility being path-dependent and its ramifications to smoke visualisation in graphic presentation of computer simulation results and to evaluation of visibility thought smoke were highlighted by [15-18]. Therefore, proper use of the correlation in conjunction with the computer simulated results deserves careful attention.

In the remaining part of this chapter, the framework for visibility and the physics of light transmission are presented. Then reviews of previous research on sign visibility and the methods of evaluating sign visibility in smoke are conducted. The limitations in the existing methods are underlined. Finally, a new approach to evaluation of sign visibility in heterogeneous smoke conditions is proposed. The advantage and the limitation of the new method are discussed, followed by recommendations for future research.

VISIBILITY AND LIGHT BEAM TRANSMISSION IN ATTENUATING MEDIA

Visibility

Things are visible or identifiable because they either emit or reflect light that stimulates the human eyes. The visibility of an object, in its common sense, describes how visible the object is. It is primarily determined by the contrast of the luminance of the object to that of its background. Contrast is defined as the ratio of luminance difference to the lower of the two luminance [19]:

$$C = \frac{|L_o - L_b|}{\min(L_o, L_b)} \tag{1}$$

where C is contrast, L_o (cdm^{-2}) is the luminance or brightness of the object and L_b is the luminance of the background. There exists a threshold value of contrast below which an object becomes indiscernible against its background. This threshold value is 0.02 [20, 21].

The contrast of an object is influenced by many factors, such as colour, size and distance, etc. [19]. In atmospheric conditions, the luminance of both the object and the background change possibly at different rates with distance due to light attenuation. The contrast may, therefore, also change with distance.

Visibility of an object in an attenuating atmospheric condition may be defined as the distance at which the contrast of the object to its background is reduced to the threshold value of 0.02. In fire safety science and engineering, visibility is defined as the maximum distance at which an object of defined size can be seen and recognised [22]. Different objects can have different visibilities under the same atmospheric condition. In fire safety engineering design and assessment, a standard reference object is warranted. The contents of a building may change during the life time of the building and they also vary from buildings to buildings. Therefore, building contents cannot be used as visual references for evacuation design. The design reference must be standardised built-in features. That design reference objects are the emergency warning lights and exit signs, since they are the standardised and fixed building features specifically for emergency evacuation purpose [23].

Re-examine Eq. (1), a question naturally arises as to what happens when the background luminance is zero. Does it mean that we have an infinite contrast? In a smoke-filled environment, the background may become completely obscured and its luminance may be close to zero. The visibility of a light emitting object will then purely depend on the luminance of the object, just like stars in a clear and light pollution free night. In this case, $L_b \approx 0$ and $\Delta L = L_o$. The visibility of the light emitting object by a naked eye is then determined by the luminance of the object [24].

Although the concepts of luminance and light intensity (radiant energy flux) are different, and the former is the modulated quantity of the latter according the human eye sensitivity to light stimulation at the visible range of light spectrum [19], the discussion on energy transmission in terms of light intensity through a grey medium is equivalent to that in terms of luminous flux. In the remaining part of this chapter, the terminologies that are familiar to fire safety scientist and engineers are used.

Transmission of Light Beam through Absorbing Media

In this subsection the basic theory that governs the attenuation of light beam in absorbing media is reviewed.

In the context of fire safety science, smoke may be treated as a semi-transparent grey medium of which the absorption coefficient (or extinction coefficient) is independent of the wave length of light [25]. According to

Bouguer-Lambert-Beer law, the decay in the transmitted intensity, or in the transmittance, follows:

$$\frac{I}{I_o} = 10^{-\int_0^S D(l)dl} = e^{-\int_0^S k(l)dl} \tag{2}$$

where I_o (Wm^{-2}) is the incident light intensity, I (Wm^{-2}) is the intensity after a distance S (m) into the attenuating medium, where $D(l)$ (m^{-1}) and $k(l)$ (m^{-1}) are commonly referred as the decadic attenuation coefficient and extinction coefficient respectively. Parameters $D(l)$ and $k(l)$ in Eq. (2) are expressed as functions of l to indicate that they may be location dependent along the path of light. If k (or D) is independent of location along the pathway, Eq. (2) becomes:

$$\frac{I}{I_o} = e^{-kS} \tag{3}$$

In the above equation, the product kS is a dimensionless parameter. It is sometimes referred as the optical depth, or optical path length [26].

The differential form of Eq. (2) reads:

$$dI = -Ik(l)dl \tag{4}$$

or

$$\frac{dI}{I} = -k(l)dl \tag{5}$$

In words, the fraction loss in the intensity of a light beam due to absorption along its path is proportional to the increment in the path length and the coefficient of the proportionality is the extinction coefficient. In the particular case where the extinction coefficient is zero, i.e., no attenuation, the light beam will maintain its intensity along its path.

Study of Sign Visibility through Smoke

It has been found that the visibility condition depends on many factors in a smoke-filled built environment, including smoke obscuration, room illumination level, light emitting or light reflecting signs, wavelength of the light, individual's visual acuity and smoke irritant content. In a study carried out by Schooley and Reagan [27], the influence of sign contrast, observer visual acuity, exposure time and threshold illumination on exit sign visibility was investigated. The variation in detectable contrast as a function of exposure time was revealed in their study. They also concluded that with sufficiently illuminated background, the contrast of signs for the intended ambient illumination, observer to sign distance and optical transmittance along the distance are the main factors to assess the legibility of an exit sign.

An extensive study to investigate the visibility of exit signs was performed at National Research Council Canada, by Clark et al. [28] who evaluated the effect of exit sign type, threshold visibility criterion and ambient smoke chamber illumination on the visibility of a range of exit signs. In their study, sixteen observers sat in a viewing booth at a distance from the sign and indicated if they could see or read the sign. Smoke was added to the chamber until the sign was obscured and the critical optical density was obtained. The data analysis indicated generally that greater smoke density was required to mask the visibility of signs with higher luminance. When ambient luminance was provided in the test chamber, sign visibility was reduced. Removing supplementary ambient illumination reduced scattered light and enabled the signs to be more visible.

The most widely referenced work is, perhaps, that by Jin in 1970s [2]. In his study, the visibilities of light reflecting and light emitting exit signs inside a smoke-filled chamber were observed by a group of participants from outside of the chamber through a glass window. The visibility measurement through smoke relied on the test subjects to determine the distance at which the object (or exit sign) was no longer visible. A set of correlations was then developed from the experimental results. Jin's work was recently verified and used by Węgrzyński and Vigne [29] in their experimental and numerical simulation study of the effect of soot yield on visibility through smoke.

It has been shown in many studies that the visibility of exit signs through smoke is related to attenuation of the brightness of the signs. The rate of attenuation, or the extinction coefficient, is determined by the soot mass concentration in smoke [2, 30, 31]. Hence, evaluation of visibility through smoke is often reduced to evaluation of smoke characteristics.

In the works of Jin [2] and other researchers [27, 28], the smoke condition in the smoke chamber can be assumed more or less uniform, i.e., a constant extinction coefficient throughout. It was found that the product of the visibility *S* and the smoke extinction coefficient *k* is almost constant for given types of smoke and signs [2]:

$$Sk = E \tag{6}$$

or

$$S = \frac{E}{k} \tag{7}$$

where *E* is visibility factor. The value of *E* depends on the threshold contrast, optical properties of smoke and optical/photometry properties of signs. Further delineation of the theoretical background of Eq. (7), or its connection with the threshold contrast theory, can be found in the article by Sychta [32]. Generally:

$$E = \begin{cases} 2\sim4 \text{ for light reflecting signs} \\ 5\sim10 \text{ for light emitting signs} \end{cases} \tag{8}$$

The ranges of variations in *E* may be related to uncertainties in lighting conditions and variability in subjects' acuity to light stimulations. Note that Eqs. (7) and (8) are valid when *k* is constant along the line of sight.

Comparing Eqs. (3) and (6) easily leads to a conclusion that *E* is in fact equivalent to a critical optical depth. Mowrer [30] commented that selecting, say $E = 8$ for light emitting signs and $E = 3$ for light-reflecting signs would suggest that "light emitting signs can be observed when the light attenuation is $I/I_o = e^{-8} = 3.35 \times 10^{-4}$, that is the transmitted light is much less than 1% of the unattenuated light intensity; whereas light-reflecting signs can be observed when the light attenuation is $I/I_o = e^{-3} = 0.05$, that is, the transmitted light is reduced to 5% of the unattenuated light level." This argument is on the basis of examining the attenuation of light intensity from standard exit signs rather than the variation in the contrast. Indeed, when the background luminance is negligibly small or nonexistence, it may suffice to examine the luminance of the light source to determine its visibility.

LIGHT RAY TRANSMISSION OF SIGNS

The above discussion on light attenuation is applicable to parallel light beams and the discussion on visibility of signs is at a specific orientation. In real life situations, light beams are generally not parallel, or their intensities vary even without attenuation. An analogy can be found in star gazing at night. Many stars and galaxies cannot be seen on Earth not only because of the light attenuation and light pollution in the atmosphere, but also because we are beaten by the sheer distances. Moreover, in the case of signage visibility, the observers of signs are most likely not directly facing the signs. Therefore, in evaluating visibility of signs, we need to take into account the factors of variability of beam intensity with distance and the sign orientation relative to the observer. For this purpose, we divide signs into two categories, namely, point source emitter and panel emitter.

Point Source Emitter

Some emergency lights, such as siren lights, warning lights or flash lights that are not modified by focusing mirrors may be treated as point sources which emit equally amount of energy in all directions. In a more technical term, the emissive power of a point source is isotropic. The light rays emitted from a point source are divergent. In fresh air without smoke, the irradiance I (light intensity or luminance) generated by the point source at a distance r can be evaluated from the inverse square law [33, 34]:

$$I = \frac{P}{4\pi r^2} \tag{9}$$

where I is light intensity or irradiance (Wm^{-2}), P is the emissive power (W) of the source and r is the distance (m). The differential form of Eq. (9) in respect to distance r is:

$$dI = -2\frac{I}{r}dr \tag{10}$$

or

$$\frac{dI}{I} = -\frac{2}{r}dr \tag{11}$$

In words, the fractional decay in irradiance due to divergence of light ray from a point source is proportional to the fractional increment in the radial distance. Parameter $2/r$ can be regarded as the divergence factor.

When the light ray passes through an absorbing medium such as smoke, a fraction of the radiant energy is absorbed and this fraction as expressed by Eq. (5) should be added to the total fractional change in irradiance. The total fractional decay is:

$$\frac{dI}{I} = -\frac{2}{r}dr - k(r)dr \tag{12}$$

A derivation of Eq. (12) is given in Appendix. The physical interpretation of Eq. (12) is that the fractional decay in irradiance is attributed to two factors: one is divergence and the other is absorption by the medium. Integrating Eq. (12) results in:

$$\ln I = -2\ln r + \ln\left(e^{-kr}\right) + \ln C \tag{13}$$

where C is the integration constant. Re-arrange the above equation into:

$$I = \frac{C}{r^2}e^{-\int_0^r k(r)dr} \tag{14}$$

When $k = 0$, Eq. (14) should be the same as Eq. (9). Therefore, we have:

$$C = \frac{P}{4\pi} \tag{15}$$

and

$$I(P,k,r) = \frac{P}{4\pi r^2} e^{-\int_0^r k(r)dr} \tag{16}$$

Equation (16) is the expression for irradiance at a distance with the consideration of attenuation by absorbing medium. Comparing Eq. (9) with Eq. (16), we find that when light rays from a point source is transmitted through an absorbing medium, its intensity is simply moderated by an exponential decay factor $e^{-\int_0^r k(r)dr}$.

At the unit radial distance of (r = 1 m) with zero attenuation (k = 0):

$$I(P,0,1) = \frac{P}{4\pi}\ (\mathrm{Wm^{-2}}) \tag{17}$$

A normalised irradiance can be introduced:

$$I^*(P,k,r) = \frac{I(P,k,r)}{I(P,0,1)} = \frac{e^{-\int_0^r k(r)dr}}{r^2} \tag{18}$$

If the extinction coefficient k is independent of the location r, Eq. (18) can be expressed as:

$$I^*(P,k,r) = \frac{e^{-kr}}{r^2} \tag{19}$$

For a light beam that is not divergent, the term $1/r^2$ drops out of Eq. (18) and it becomes:

$$I^*(P,k,r) = e^{-\int_0^r k(r)dr} \tag{20}$$

This is simply the transmittance equation for parallel light beams through absorbing medium, i.e., Eq. (2). If k is independent of r, we have:

$$I^*(P,k,r) = e^{-kr} \tag{21}$$

Log-log plots of Eqs. (19) and (21) for a range of extinction coefficient values are presented in Figure 1, together with the two lines indicating the threshold criteria for light reflecting and light emitting signs respectively.

Significant differences can be discerned between Figure 1(a) and Figure 1(b). Under the zero extinction coefficient condition, i.e., no smoke, the parallel beam maintains its intensity indefinitely. Whereas, the intensity of light from a point source decreases dramatically with distance. The presence of smoke, or non-zero extinction coefficient only makes the decrease go faster. In general, Eqs. (18) and (19) produce more conservative estimate of reduction in irradiance than Eqs. (20) and (21).

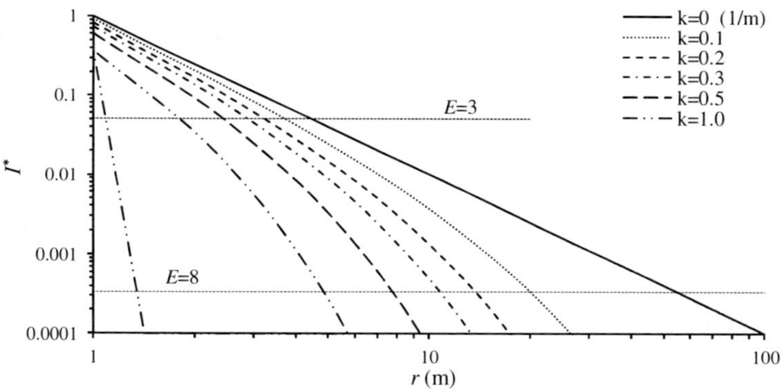

(a) Transmittance of diverging light from a point source [Eq. (19)].

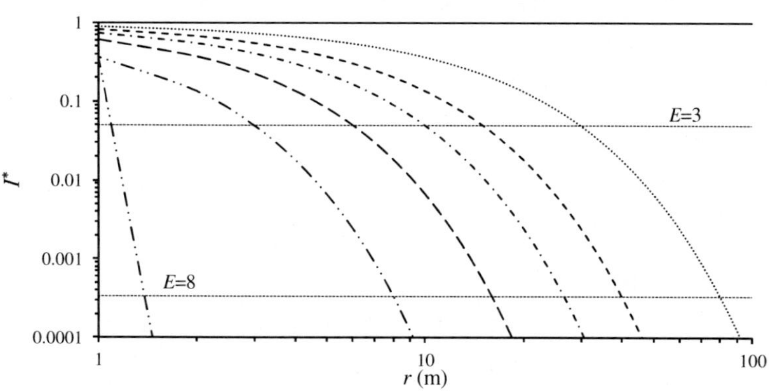

(b) Transmittance of parallel light beam [Eq. (21)].

Figure 1. Transmittance of light from two kind of light rays and under a range of extinction coefficients.

Panel Source Emitter

Most exit signs are more like panels than point sources in that their visibility depends not only on the distance but also on the angle of observation. The analysis of their visibility is slightly more intricate.

An exit sign may be treated as light emitting panel of diffuse surface A_1. A human eye is represented by a small element dA_2 at a distance r. See Figure 2.

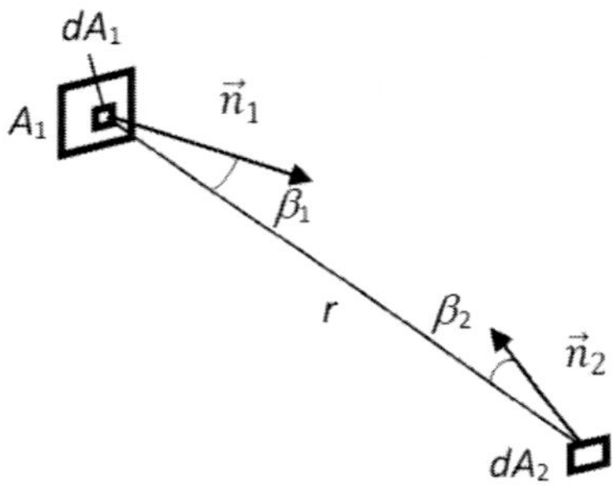

Figure 2. Geometry between two surfaces.

The luminous flux from A_1 to dA_2 is equivalent of irradiance received by dA_2 from A_1 and is related to view factor (or configuration factor) F [35]:

$$F = \int_{A_1} \frac{\cos\beta_1 \cos\beta_2}{\pi r^2} dA_1 \tag{22}$$

where dA_1 is a small element in A_1, r is the distance between dA_1 and dA_2, β_i (i = 1, 2) is the angle between the line of sight and the norm $\vec{n}_i$ of the corresponding surface.

For $\sqrt{A_1} \ll r$, the variation in β_1 is negligibly small and $\cos\beta_1$ is approximately constant over the integration region A_1. Likewise, for an eye looking at the sign A_1, $\beta_2 \approx 0$. Integrating Eq. (22) results in:

$$F = \frac{\cos\beta_1}{\pi r^2} A_1 \tag{23}$$

If the total power of the sign is P, its emissive power is P/A_1 and the irradiance at the location of the eye is:

$$I = F\frac{P}{A_1} \tag{24}$$

Substituting Eq. (23) into Eq. (24) yields:

$$I = \frac{\cos\beta_1}{\pi r^2}P \tag{25}$$

For simplicity, the subscript 1 in the angle β_1 is dropped and we have:

$$I = \frac{\cos\beta}{\pi r^2}P \tag{26}$$

Note that the divergence factor $1/r^2$ is incorporated into Eq. (26) which says that the luminance of a panel like exit sign at a considerable large distance away is proportional to its power (luminosity) and orientation relative to the observer and inversely proportional to the square of distance.

Similar to the relationship between Eqs. (9) and (16), when the attenuation by an absorbing medium such as smoke is considered, the irradiance at a distance away is:

$$I(P,\beta,k,r) = \frac{\cos\beta}{\pi r^2}Pe^{-\int_0^r k(r)dr} \tag{27}$$

It is noted that the difference between Eq. (27) and Eq. (16) is the $\cos\beta$ function. In words, the transmittance of light from a diffuse surface sign is similar to the transmittance of the light from a point source but with a moderation by the cosine of the angle between the line of sight and the norm of the sign surface.

Again, we use the irradiance at zero angle (i.e., normal to the sign panel), zero extinction coefficient (i.e., no smoke) and unit distance:

$$I(P,0,0,1) = \frac{P}{\pi} \tag{28}$$

as a reference value and define the normalised quantity:

$$I^*(P,\beta,r,k) = \frac{I(P,\beta,k,r)}{I(P,0,0,1)} = \frac{\cos\beta}{r^2} e^{-\int_0^r k(r)dr} \tag{29}$$

For homogeneously distributed extinction coefficient, Eq. (29) is simplified into:

$$I^*(P,\beta,r,k) = \frac{I(P,\beta,k,r)}{I(P,0,0,1)} = \frac{\cos\beta}{r^2} e^{-kr} \tag{30}$$

Equation (30) is plotted in Figure 3 for a range of constant extinction coefficient values to demonstrate the dependence of the normalised irradiance on viewing angle β as well as distance r.

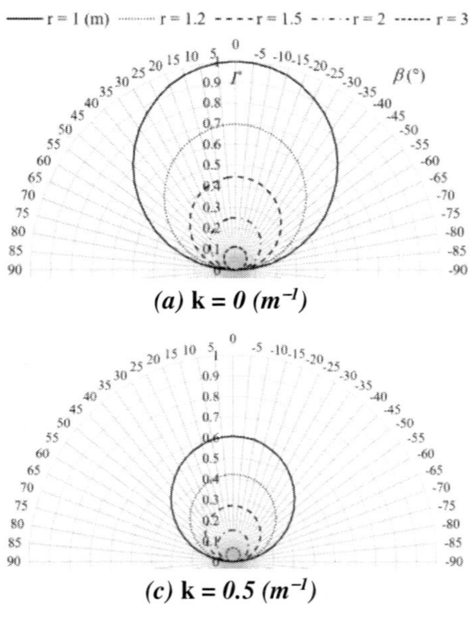

Figure 3. Variation of the normalised irradiance received from a diffusive panel as function of viewing angle, distance and smoke extinction coefficient.

Evaluation of Visibility of Signs from Computer Simulation Results

Computational fluid dynamics models, such as Fire Dynamics Simulator (FDS) [36], are capable of predicting smoke properties, more specifically smoke concentration and extinction coefficient *k*, distributions over the computation domain. This allows the evaluation of visibility through smoke for fire safety engineering design and event re-construction purposes. Various approaches may be adopted for the evaluation. One of them is to assess visibility at any given location in the simulated environment. Another is to assess the visibility of designed objects, namely exit signs in fire scenarios. These two approaches are discussed in details in the following subsections.

Evaluation of Visibility from Evacuee's Perspective

The first approach gives an evacuee's perspective, i.e., how far an evacuee can see in a fire scenario. Examples of the first approach can be found in FDS [9] which incorporates Jin's correlation [Eq. (7)] to assess the local visibility. However, a large degree of uncertainty may be casted on this approach. There are two contributors to the uncertainty. Firstly, visibility is an extensive parameter. It depends on the smoke condition along the line of sight between human eye and the object to be seen. On the other hand, smoke extinction coefficient *k* is an intensive parameter, therefore a correlation of the form in Eq. (7) is valid only if *k* is homogeneously distributed along the line of sight. To address this problem Kang and Macdonald [16] proposed a ray-tracing method in which the decay of light intensity is attributed to the integrated attenuation along the line of sight and there is a threshold intensity below which the object will not be seen. The corresponding distance is the visibility.

A question relating to the second contributor to the uncertainty then arises as to how to determine the threshold intensity. Kang [37] employed the concept of opacity defined as:

$$O_p = 1 - \frac{I}{I_o} \tag{31}$$

in a graphical approach to describing visibilities of objects in smoke scene. However, different objects may have different brightness (or intensity) I_o under the same lighting condition. Setting a universal opacity could result in different threshold intensity.

A similar ray-tracing method was described by Husted et al. [15] in their exercise to evaluate visibility through inhomogeneous smoke. Their selection of the threshold value is essentially based on setting the transmittance value to 0.1 and the corresponding distance is the visibility. This value seems very strenuous in comparison to the threshold transmittance value deduced from Jin's study of sign visibility. See the discussion at the end of Section 0 of the current chapter.

Evaluation of Visibility of Standard Object – Exit Signs

With the consideration that a standard reference object is needed for the evaluation of visibility and the most critical reference object in the emergency evacuation is the exit sign, He [18] extended Kang and Macdonald's [16] ray-tracing technique by incorporating Jin's [2] correlation and visibility factor as the threshold value for evaluating visibility. In He's work, the constant extinction coefficient *k* in Jin's correlation, i.e., Eq. (7), is replaced with the average extinction coefficient over the distance along the line of sight between the observer and the exit sign. A parameter of virtual visibility is defined as:

$$S_v = \frac{E}{K_{aL}} \tag{32}$$

where K_{aL} (m^{-1}) is the average extinction coefficient along the line of sight:

$$K_{aL} = \frac{1}{L}\int_0^L k(l)dl \tag{33}$$

and L is the length of the line of sight between the eye of the subject and the sign. In words, virtual visibility is the visible distance that is evaluated from the average extinction coefficient along the line of sight of given distance L. If the virtual visibility S_v is greater than the distance L, then the sign is visible and otherwise, invisible.

Evaluation of Visibility of Exit Signs Over a Domain

The original application of the aforementioned method, as demonstrated in He's article [18], was still limited to the evacuees' perspective in that the line of sight is between an individual evacuee and an exit sign. To apply this method in a broader perspective for engineering design practice, it is proposed here to employ the concept of visibility catchment area of signs [38] and carry out the field evaluation, i.e., instead of evaluating the visibility of a given sign to an individual, the field in which a given sign is visible to any individual within it can be estimated. The process is delineated below:

1. In a given compartment, determine the locations and the nature (point source or panel source, light reflecting or light emitting) of the warning light or exit sign.
2. Select the appropriate value for the visibility factor E.
3. Originating from the location of the sign, set a number of rays over an angular range (may be solid angular range if a three-dimensional evaluation is necessary).
4. Perform the integration $e^{-\int_0^r k(r)dr}$ along the rays.
5. Use Eq. (18) to evaluate I^* for point source or Eq. (29) for panel sign.

6. Along each ray, when I^* attains the threshold value of e^{-E}, the corresponding r value defines the boundary of the visible domain of the given sign in the given ray direction. Note that the rays may be bounded by the physical boundaries of the compartment.
7. The catchment area of the sign is defined by all the rays of finite lengths.

The above described method has the advantage of not having to determine individual evacuees' positions in the domain of concern. It can simply generate a map of visible zones for the allocated signs. If any region of the domain is not covered by any visible zones, then it is a potential untenable region.

DISCUSSION

In many engineering design cases, the evacuation paths may constrain the exit sign viewing angle to a small range. For example, when evacuees are travelling along a corridor, the viewing angle of an exit sign along the corridor may not vary much. In other built environments, such as large open office spaces, industrial plants, movie theatres and indoor sports stadium, the occupants may be situated in locations not in the right front of exit signs near them. In these cases, consideration of view angles in the evaluation of visibility of signs is necessary.

By taking into account the divergence of light rays, Eqs. (18) and (29) results in more conservative estimate, from the fire safety point of view, of visibility than the theory for non-diverging beams. Equation (29) is valid for distances much greater than the size of the sign panel. For distances comparable to the size of the sign, one needs to resort to more accurate evaluation of the view factor. However, if an evacuee is at a distance comparable to the size of the sign, s/he may be practically at the exit.

It should be pointed out that despite their appearances, the quantities evaluated from Eqs. (18), (19), (29) and (30) are dimensionless.

Although not expressed explicitly, the previous empirical studies and the derived correlations may have already included the divergence effect of the transmission of lights from signs. A careful analysis is warranted to siphon out this effect amidst the collected data.

Light extinction is a combined effect of light absorption and scattering [17, 39]. The scattered light from an ambient source may be seen as reflection from soot particles which obscures the vision of the signs. The method outlined in the foregoing subsection does not take into account the effect of ambient illumination on visibility. This effect has been shown to be significant [28] and needs to be considered in future studies.

The discussions in most part of this chapter have been conducted using the terminologies relating to radiant energy transfer. They can be translated into the corresponding terminologies in photometry by weighting the energy spectrum with the luminous efficacy curve [19].

The visibility factor for determining the threshold transmittance is a rather empirical parameter, the selection of which for visibility assessment involves a considerable degree of uncertainty. In-depth literature review and more experimental studies are required to introduce an appropriate photometric parameter with a more definitive threshold value of transmittance. If the large variation in visibilities is related to variability in human eyes acuity to light, then it is recommended that the uncertainty may be dealt with in human response analysis such as those conducted by Occhialini et al. [13].

Advancement is seen recently in signage visibility analysis using building information modelling enabled virtual reality environments [8]. The proposed method in the current chapter can be incorporated into the signage visibility analysis under smoky conditions.

CONCLUSION

Over the years, *smoke visibility* is regarded as a property of smoke in many reference books and publications. This chapter highlights that visibility is an extensive parameter. It is not a property of smoke, nor is it dependent on the local smoke condition alone. It is related to the distribution of physical properties of smoke along the line of sight as well as the object that is to be visualised. Perhaps, a more appropriate terminology is *visibility through smoke*. Moreover, visibility through smoke is also dependent on the relative position of the human subject to the observed object. In the case of exit sign panels, it is also dependent on the orientation of the panel relative to the observer.

Based on a review of the physics of light transmission and empirical studies of sign visibility, a new approach to evaluation of sign visibility for fire safety engineering design and assessment is proposed. This approach incorporates the divergence effect in light transmission from various sources, the orientation of signs and the heterogeneity in smoke property distribution within the domain of concern.

The proposed approach is a step forward towards the sophistication in evaluating sign visibility through smoke. Nonetheless, further refinement is needed to include the effect of ambient illumination. If the method is to be used to evaluate the visibility of non-standard objects, then the luminosity of the objects should be considered.

APPENDIX

The principle of conservation of energy demands that the net change in radiant energy flow across the body of finite volume is due to light absorption within the volume enclosed by the area. For a light absorbing spherical space where an isotropic light source is located at the centre, we may take a thin spherical shell with the thickness of dr as the control volume. The differential form of the conservation of energy can be expressed as, by incorporating Bouguer-Lambert-Beer law:

$$d(IA) = -Ik(r)Adr \tag{34}$$

where A is the area of the sphere. Expand and re-arrange the above equation. We have:

$$\frac{dI}{I} = -\frac{dA}{A} - k(r)dr \tag{35}$$

For spherical surface:

$$\frac{dA}{A} = \frac{2}{r}dr \tag{36}$$

Therefore, Eq. (12).

ACKNOWLEDGMENTS

The author is indebted to Dr Wallace X. Zhong for his valuable comment and suggestions on drafting of this chapter.

REFERENCES

[1] Yamada, T., & Y. Akizuki. (2016). Visibility and Human Behaviour in Fire Smoke. In *SFPE Handbook of Fire Protection Engineering*, edited by M. J. Hurley, 2181-2206. New York: Springer.

[2] Jin, T. (1978). Visibility through Fire Smoke. *Journal of Fire and Flammability* 9:135-157.

[3] Ronchi, E., Gwynne, S., & Purser, D. A. (2011). The impact of default settings on evacuation model results: a study of visibility conditions vs occupant walking speeds. In: *Advanced Research Workshop Evacuation and Human Behaviour in Emergency Situations EVAC11*, Santander, 81-95.

[4] Cao, S., Fu, L., Wang, P., Zeng, G., & Song, W. (2018). Experimental and modeling study on evacuation under good and limited visibility in a supermarket. *Fire safety journal*, 102, 27-36.

[5] Kuligowski, E. D. (2016). Human Behavior in Fire. In *SFPE Handbook of Fire Protection Engineering*, edited by M. J. Hurley, 2070-2114. New York: Springer.

[6] Bryan, J. L. (1983). *Implications for Codes and Behavior Models from Analysis of Behavior Response Patterns as Selected from Project People and Project People II Study Programs*. The US Department of Commerce and the U.S. Department of Health and Human Services.

[7] ICC. (2005). *International Fire Engineering Guidelines*. Canberra, Australia: International Code Council, National Research Council of Canada, New Zealand Department of Building and Housing, Australian Building Codes Board.

[8] Motamedi, A., Wang, Z., Yabuki, N., Fukuda, T., & Michikawa, T. (2017). Signage visibility analysis and optimization system using BIM-enabled virtual reality (VR) environments. *Advanced Engineering Informatics*, 32, 248-262.

[9] McGrattan, K., & Miles, S. (2016). Modeling fires using computational fluid dynamics (CFD). *In SFPE Handbook of Fire Protection Engineering,* 1034-1065. Springer, New York, NY.

[10] Simcox, S., Wilkes, N. S., and Jones, I. P. (1992). Computer simulation of the flows of hot gases from the fire at King's Cross underground station. *Fire Safety Journal* 18 (1):49-73.

[11] Bryner, N., Madrzykowski, D., & Grosshandler, W. (2007, September). *Reconstructing the station nightclub fire–computer modeling of the fire growth and spread.* In International interflam conference, 11th proceedings (pp. 3-5).

[12] Chi, J. H. (2012). Metallographic Analysis and Fire Dynamics Simulation for Electrical Fire Scene Reconstruction. *Journal of Forensic Science* 57 (1):246-249.

[13] Occhialini, M., Bernardini, G., Ferracuti, F., Iarlori, S., D'Orazio, M., & Longhi, S. (2016). Fire exit signs: The use of neurological activity analysis for quantitative evaluations on their perceptiveness in a virtual environment. *Fire Safety Journal*, 82, 63-75.

[14] Chen, J., Wang, J., Wang, B., Liu, R., & Wang, Q. (2018). An experimental study of visibility effect on evacuation speed on stairs. *Fire safety journal*, 96, 189-202.

[15] Husted, B. P., Carlsson, J., & Göransson, U. (2004). *Visibility through inhomogeneous smoke using CFD*. Interflam 2004, 10th Proceedings.

[16] Kang, K., & Macdonald, H. M. (2005). Modeling Smoke Visibility in CFD. *Eighth International Symposium on Fire Safety Science*, Beijing, China, September 18-23.

[17] Rubini, P. A., Q. Zhang, and Moss, J. B. (2007). *Simulation of Visibility in Smoke Laden Environments*. InterFlam 2007, 11th International Conference on Fire Science and Engineering, University of Hull, London, UK, 3-5 September, 2007.

[18] He, Y. (2009). Evaluating visibility using FDS modelling result. FSE09: *Fire Safety Engineering International Conference: Charting the Course*, Melbourne, 18–19 March.

[19] Szokolay, S. V. (2014). *Introduction to Architectural Science: The Basis of Sustainable Design*. 3rd ed: Routledge.

[20] Blackwell, H. R. (1946). "Contrast Thresholds of the Human Eye." *Journal of the Optical Society of America* 36 (11):624-643.

[21] McCartney, E. J. (1976). *Optics of the Atmosphere*. New York: Wiley and Sons.

[22] ISO 13943. (2000). *Fire Safety – Vocabulary*. International Organization for Standardization.

[23] AS/NZS-2293.1. (2018). *Emergency lighting and exit signs for buildings - System design, installation and operation*. Sydney: Standard Australia.

[24] Crumey, A. (2014). Human contrast threshold and astronomical visibility. *Monthly Notices of Royal Astronomical Society* 442:2600–2619.

[25] Lattimer, B. Y. (2016). Heat Transfer from Fires to Surfaces. In *SFPE Handbook of Fire Protection Engineering*, edited by M. J. Hurley, 745-798. New York: Springer.

[26] Lautenberger, C., Tien, C. L., Lee, K. Y., Stretton, A. J. (2016). Radiation Heat Transfer. In *The SFPE Handbook of Fire Protection Engineering*, edited by M. J. Hurley, 102. New York: Springer.

[27] Schooley, L. C., & Reagan, J. A. (1980). Visibility and Legibility of Exit Signs. *Journal of the Illuminating Engineering Society* 10:24-28.

[28] Clark, F. R., Rea, M. S., & Ouellette, M. J. (1985). *Visibility of exit signs through smoke*. National Research Council Canada, Division of Building Research.

[29] Węgrzyński, W., & Vigne, G. (2017). Experimental and numerical evaluation of the influence of the soot yield on the visibility in smoke in CFD analysis. *Fire Safety Journal*, 91, 389-398.

[30] Mowrer, F. W. (2016). Enclosure Smoke Filling and Fire-Generated Environmental Conditions. In *SFPE Handbook of Fire Protection Engineering*, edited by M. J. Hurley, 1066-1101. New York: Springer.

[31] Mulholland, G. W., & Croarkin, C. (2000). Specific Extinction Coefficient of Flame Generated Smoke. *Fire and Materials* 24 (5):209-252.

[32] Sychta, Z. (1997). Contrast Attenuation Coefficient as a Parameter Enabling Determination of Range of Visibility in Smoke. *Fire and Materials* 21 (5):205-211.

[33] Greenland, J. (1998). *Foundations of architectural science: heat, light, sound*. Sydney: University of Technology Sydney, Faculty of Design, Architecture and Building.

[34] Drysdale, D. D. (2011). *An Introduction to Fire Dynamics*. 3rd ed. Chichester, UK.: John Wiley and Sons.

[35] Siegel, R., & Howell, J. R. (2002). *Thermal Radiation Heat Transfer*. 4th ed: Taylor & Francis.

[36] McGrattan, K., R. McDermott, S. Hostikka, J. Floyd, C. Weinschenk, and Overholt, K. (2018). Fire Dynamics Simulator User's Guide. *NIST Special Publication 1019*, National Institute of Standards and Technology.

[37] Kang, K. (2007). A smoke model and its application for smoke management in an underground mass transit station. *Fire Safety Journal* 42:218–231.

[38] Filippidis, L., E. R. Galea, P. Lawrence, and Gwynne, S. (2001). *Visibility Catchment Area of Exits and Signs*. Interflam 2001, Edinburgh, Scotland, 17 - 19th September.

[39] Mulholland, G. W. (2008). Smoke Production and Properties. In *The SFPE Handbook of Fire Protection Engineering*, edited by P. J. DiNenno, 2.291-2.302. Boston: Society of Fire Protection Engineers.

BIOGRAPHICAL SKETCH

Yaping He

Affiliation: School of computing, Engineering and Mathematics, Western Sydney University

Education: PhD., M.Eng.Sci., B.E.

Business Address:
Locked Bag 1797
Penrith
NSW 2751
Australia

Research and Professional Experience:

Dr He has research interest in the areas of fire safety science, fire dynamics, risk analysis, performance based design. Contributed to the developments of *Australian Fire Safety Engineering Guidelines* and *International Fire Engineering Guidelines.*

Professional Appointments:

- School of computing, Engineering and Mathematics, Western Sydney University (2004 - present), Senior Lecturer
- Scientific Services Laboratory (1998 – 2004) Senior Fire Safety Engineer
- Victoria University (1994-1998), Senior Research Associate

Honors:

Recipient of 2011 Jack Bono Award by Society of Fire Protection Engineers

Publications from the Last 3 Years:

1. Eftekharian, E., Ghodrat, M., He, Y., Ong, R. H, Kwok, K. C. S., Zhao, M., 2019, Numerical analysis of wind velocity effects on fire-wind enhancement, *International Journal of Heat and Fluid Flow*, 80 (2019) 108471.
2. Eftekharian, E., Ghodrat, M., He, Y., Ong, R. H., Kwok, K. C. S., Zhao, M., Samali, B. 2019. Investigation of terrain slope effects on wind enhancement by a line source fire, *Case Studies in Thermal Engineering*, 14, 100467.
3. Douglas, G. B., He, Y. 2019. Design Bushfire Selection for Bushfire Protection in Adaptation to Global Warming. *Frontiers in Mechanical Engineering,* 5, Article 27.
4. Eftekharian, E., He, Y., Kwok, K. C. S., Ong, R. H., Yuan, J., 2019, Investigation of fire-driven cross-wind velocity enhancement, *International Journal of Thermal Sciences*, 141: 84–95.
5. Ouyang, D., He, Y., Weng, J., Liu, J., Chen, M., Wang, J., 2019, Influence of low temperature conditions on lithium-ion batteries and the application of an insulation material, *RSC Advances*, 9: 9053–9066.
6. Dang, X., He, Y., Wang, J., 2018, Experimental and numerical study on burning behaviour of pool fire with different lip heights, *Fire Safety Science*, (Chinese) 27(4), 213–221.
7. Wang, X., He, Y., Zhou, T., Chen, Q., Ding, C., Wang, J., 2018, Experimental study on fire behaviors of kerosene-additives blends, *Fire Technology*, 54: 1841–1869.
8. Chen, Q., He, Y., Wang, X., Zhou, T., Ding, C., Wang, J., 2018, Flash point and fire hazard analysis of three organic fuels and their

aqueous solutions at low pressures, *Fire and Materials*, 42:805–815.

9. Wei, R., He Y., Zhang, Z., He, J., Yuen, R., Wang, J., 2018, Effect of different humectants on the thermal stability and fire hazard of nitrocellulose, *Journal of Thermal Analysis and Calorimetry*, 133:1291–1307.
10. Yuan, C., He, Y., Feng, Y., Wang, P., 2018, Fire hazards in heritage villages: A case study on Dangjia Village in China, *International Journal of Disaster Risk Reduction*, 28, 748–757.
11. Ouyang, D., He, Y., Chen, M., Liu, J., Wang, J., 2018, Experimental study on the thermal behaviors of lithium-ion batteries under discharge and overcharge conditions, *Journal of Thermal Analysis and Calorimetry*, 132(1), 65-75.
12. Lin, X., He, Y., Jiang, W., Liu, J., Chen, M., Yao, W., Ma, P., Ding, C., Wang, J., 2017, Prediction of heat release rate of porous fuel based on profile burning surface, *Journal of Thermal Analysis and Calorimetry*, 130(3), 2215–2225.
13. Wang, S., He, Y., Yang, H., Wang, K., Wang, J., 2017, Video smoke detection using shape, color and dynamic features, *Journal of Intelligent and Fuzzy Systems*, 33(1), 305 – 313.
14. Tao, C., He, Y., Zhuang, Y., Qian, Y., Cheng, X. and Wang, X., 2017, The Investigation of Flame Length of Buoyancy-Controlled Gas Fire Bounded by Wall and Ceiling, *Applied Thermal Engineering*, 127, 1172–1183.
15. Chen, M., Liu, J., He, Y., Yuen, R., Wang, J., 2017, Study of the fire hazards of lithium-ion batteries at different pressures, *Applied Thermal Engineering*, 125, 1061-1074.
16. He, Y., He, Y., Liu, J., Li, P., Chen, M., Wei, R., Wang, J., 2017, Experimental study on the thermal decomposition and combustion characteristics of nitrocellulose with different alcohol humectants. *Journal of Hazardous Materials*, 340, 202–212.
17. He, Y., Park, L., 2017, A statistical analysis of occurrence and association between structural fire hazards in heritage housing, *Fire Safety Journal*, 90, 169-180.

18. Wei, R., He, Y., Liu, J., He, Y., Mi, W., Yuen, R., Wang, J., 2017, Experimental Study on the Fire Properties of Nitrocellulose with Different Structures, *Materials*, 10, 316.
19. Zhou, T., He, Y., Lin, X., Wang, X., Wang, J., 2017, Influence of constraint effect of sidewall on maximum smoke temperature distribution under a tunnel ceiling, *Applied Thermal Engineering*, 112, 932–941.

In: Critical Issues and Analysis … ISBN: 978-1-53618-738-0
Editors: A. Cablé and B. Manescau

Chapter 3

STUDY OF THE FIRE BEHAVIOR OF TWO COMPOSITE MATERIALS: CARBON-PHENOLIC AND CARBON-PEKK

Brady Manescau[1*], Khaled Chetehouna[1], Nathan Grange[1], Nicolas Gascoin[1], Samuel Senave[2] and Isabelle Reynaud[2]

[1]INSA Centre Val de Loire, University of Orléans, Bourges, France
[2]DAHER, St. Julien de Chédon, France

ABSTRACT

Over the past 50 years, the aerospace industry has made extraordinary progress in optimizing the materials used in the design of various flying vehicles. In particular, the research and development efforts have highlighted the major advantages of using composite materials in the manufacturing of aircraft and space rockets. In addition to their lightness, a review by Kumar et al. (2019) shows that composite materials have remarkable structural and mechanical properties, and that the

[*] Corresponding Author's Email: brady.manescau@insa-cvl.fr.

implementation of new matrix manufacturing combinations can further improve these properties in order to increase the strength / weight ratio. However, if composite materials guarantee good mechanical properties, they exhibit specific and variable behaviors when subjected to a fire. With this in mind, before being used in the design of aircrafts, composite materials must be subjected to different fire resistance certification tests carried out on different scales. In this chapter, the fire behavior of two carbon reinforced composite materials (carbon-phenolic and carbon-PEKK) was studied on a medium and large scale. First, the gaseous species produced during the thermal degradation of both composite materials was evaluated by means of measurement campaigns. The tests were carried out using a calorimeter cone and a NexGen burner. Mass loss, rear face temperature measurements as well as a characterization of gas emissions were carried out. These made it possible to make a detailed description of the two composite materials studied and to compare their performances. Gaseous emissions were measured using analytical pyrolysis (heated platinum filament) coupled to a gas chromatograph and a mass spectrometer (Py-GC - MS). The mass loss was measured using a load cell and the rear face temperatures were obtained using a thermal camera and type K thermocouples. From pyrolysis gas analyzes, the lower limit d flammability (LII) was obtained. By considering rear face temperature and mass loss, it has been found that the carbon-PEKK-based composite material has better performance compared to the carbon-phenolic-based composite material. This superiority is assessed by demonstrating better thermal stability and low combustibility of the carbon-PEKK-based composite material.

Keywords: carbon phenolic, carbon-PEKK, NexGen burner, thermal degradation, gaseous emissions, medium and large scale

INTRODUCTION

Since the industrial revolution, engineers have been constantly looking for solutions to optimize the energy performance of different means of transport. Among these means of mobility, there are the flying machines characterizing the aeronautical industry and in this lattest, composite materials profoundly improved the design of aircraft due to their main advantages in weight reduction, [1] and strong mechanical properties [2] allowing significant cost reductions. Nevertheless, these materials also present many advantages in every modern transportation systems, such as trains [3], planes [4], automobiles [5] and even bicycles [6]. However, the quest of utilizing composite materials in modern transport system is still in its infancy due to being vulnerable when exposed to fire or even high temperatures. Unfortunately, during their thermal degradation, composite materials serve as a fuel and also produce toxic fumes resulting in an explosive atmosphere. Thus, fire spread might continue even after the original source of fire is extinguished or depleted [7].

To address these issues, various experiments involving composite materials were performed at different scales to understand, to predict and to certify their behavior when they are exposed to fire [8]. Particularly, in the aerospace industry, materials have to achieve standard fire tests at large scale; following rules such as ISO 2685 [9] or FAA AC 20-135 [10]. These tests, which involve flat panels or even complex airplane parts, aim at ensuring that every constitutive element of the aircraft retain their function for the longest time possible to guarantee the safety of passengers or carried goods.

Indeed, composite materials, in addition to their mechanical duty, are used to provide a fire protection in aeronautics. The constant evolution of fire safety standards has raised the bars for composite materials. Thus, to guarantee good protection from fire, these materials must sustain fire as long as possible with low amount of fume production to allow passenger to evacuate safely.

In this chapter, the fire behavior is experimentally investigated at two different scales. At medimu scale (samples about 10 cm×10cm) a cone calorimeter (radiative heating) device is used, which is frequently employed to assess the fire behavior of composite materials estimating heat release rate [11], flammability [12] or fire performances [13]. Moreover, this kind of test is considered as a standard test for materials [14] providing a realistic degradation environment (heat flux ranges from 30 to 120 kW/m^2). It provides Heat Release Rate (HRR), surface temperature and Mass Loss Rate (MLR), which are the required parameters to evaluate the fire behavior of composite materials; for example, MLR is one of the most important factors as it is directly linked to the pyrolysis rate, and represents the initial cause of the combustion process [15]. At large scale, a NexGen Burner (designed by the FAA) as per the aeronautical standards with a kerosene flame is used [9, 10] to test samples (about 50 cm×50 cm) which are representatives of parts of aircraft structures. By comparing the results obtained from these two scales, the relevance of tested materials at different scales are evaluated and the degradation behavior of tested specimens using the two different heat sources is compared (i.e., radiative heat flux and kerosene flame). Therefore, to contribute to the ongoing work on the assessment of fire behavior of composite materials at the above-mentioned two different scales, the fire behavior of two carbon-reinforced composites cured from pre-impregnated fibers are studied. The first is a thermosetting phenolic, which finds multiple applications in the industry for thermal protection, especially in the aerospace industry [7]; moreover, it is also used for automotive components or fire barriers in offshore oil platform [16]. Under fire conditions, carbon-phenolic composites are characterized by a high yield of char (up to 50% of the initial mass) and a low thermal conductivity [17] involving a slow ignition time and an easy extinguishment [18] due to moderate number of volatiles yield and reduced flammability [19]. Previous work done in this domain demonstrates that degradation of the carbon-phenolic is driven by two reactions under inert atmosphere with respective mass loss of approximately 20%, and by four reactions under oxidative atmosphere with complete depletion of materials at the end of test [20]. The thermal characterization of the carbon-phenolic composite also illustrates that this material has a high specific heat and reduced thermal conductivity at high temperature enhancing its ability to withstand fire [17].

The second material will be the carbon-PEKK thermoplastic. This thermoplastic polymer which is known for its high mechanical performance was originally designed for external structure and cabin interiors in aerospace as well as automotive industry [19, 20, 21]. The monomer of PEKK resin is characterized by three aromatic groups in the main chain with ether and ketone linkages, providing high thermal stability, with a glass transition and a melting temperature of, Tg = 156°C and a melting temperature (Tm) equal to 338°C respectively [22]. Moreover, the high yield of stable carbonaceous char during the decomposition process (~60% of initial mass) leads to a reduction in volatiles during combustion and display superior flammability resistance [23, 24]. Thermogravimetric analysis at small scale demonstrates that the resin degradation starts at about 500°C with a remaining mass loss of less than 20%, whereas under oxidative atmosphere, no residual amount of materials was found and it presents three global reactions that are associated with the resin pyrolysis, char and carbon-fiber oxidation [22]. The carbon-PEKK decomposition is driven by one major reaction under inert atmosphere (resin pyrolysis) and three under oxidative atmospheres (resin pyrolysis, char and carbon-fibers oxidation). Moreover, the measurement of thermal properties confirms the higher thermal stability of the carbon-PEKK [17].

The goal of this chapter will be to give description the thermal properties of the two composite material presented before based on their fire behavior. For this, experiments will be carried at medium scale and large scale using a cone calorimeter and a NexGen burner. Thanks to the data get at these two scales, a comparison of the phenomema at these two scale will be done in order to highlight whether perform experiment test at medium scale is accurate. In a second time, using the experimental observations and results, such as the backward face temperatures, mass loss ratio, gaseous emissions and heat release rates, it will be possible to highlight which of the two composite materials has the best thermal stability.

EXPERIMENTAL DEVICES

Medium Scale Set Up

At medium scale, the experiments were carried out using a cone calorimeter based on the ISO 5660-1: 2002 standard [14]. A sample with an exposed surface of 88.4 cm² was placed at 35 mm from a heat source under a constant heat flux of 100 kW/m² and measurements were recorded. The samples were placed horizontally and vertically to evaluate the result for both configurations (cf. Figure 1). To experiments are performed twice to ensure the reliability of the data obtained in this study. Moreover, the Mass

Loss Rate (MLR), the Heat Release Rate (HRR) and the Backward Face Temperature (BFT) are measured while performing the experiments. In the vertical configuration the measure of the BFT is made using a Flir A600 infrared thermography apparatus which allowed to measure the temperature field on a zone of located at the center of the backward face of the sample covered by a graphite layer (with a known emissivity of). In the horizontal configuration, the MLR is measured using a balance situated under the sample holder. The HRR is determined using the oxygen depletion principle [25] based on the concentration of O2, CO and CO2. To this end, smoke is absorbed at 24 l/s with a gas sampling of the combustion products at 58.3 ml/s. The HRR is estimated with a global uncertainty of 5%, and with a confidence level of 95%.

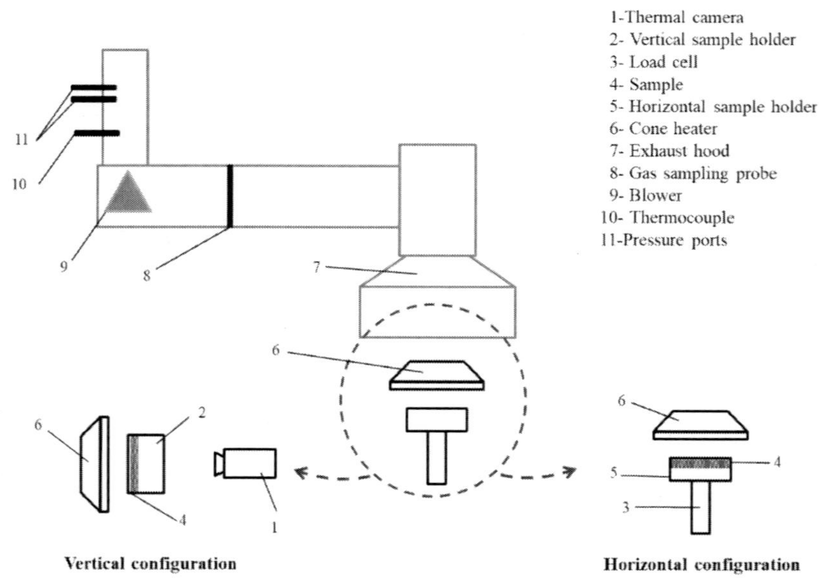

Figure 1. Overview of the cone calorimeter experiments.

Large Scale Set Up

For the large scale experiments, the set up used was the NexGen burner (see Figure 2). These are conducted in accordance to the fire standards accepted in the aerospace industry (ISO 2685 and FAA AC20-135), using a next generation fire burner (NexGen Burner,) designed by the FAA, (see Figure 2). This kind of test bench has been designed for certifications and is a representative of realistic fire conditions, close to those encountered in real-life accidents, particularly in areas around aircraft engines. As mentioned in the standards, the samples must withstand a fire exposition of 15 minutes to a kerosene-air diffusion flame without igniting the material or self-ignition in the unexposed areas. The impinging flame must ensure a heat flux of 116 kW/m^2 ± 10kw/m^2 and a mean flame temperature of 1100°C ± 80°C [10].

These fire conditions are generated using a manually operated air supply circuit which is providing pressurezied air ranging from 1.90 bar to 4.5 bar generated with a compressor to the burner. To ensure identical conditions for each run, the inlet temperature in the burner is stabilized between 5 and 15°C. The air is passed through a water exchanger allowing it to get cooled and then it is circulated in a coil stored in a freezer to stabilize the temperature, the air is then passed through a sonic throat providing a constant flow at atmospheric pressure at the outlet with limited pressure variations. The fuel used is Jet A-1 kerosene, which is vaporized into the burner cone using a Delavan injector with a temperature between 0 and 10°C. The injector ensures an approximate output flow of kerosene of 2.5 GPH (2.10x10-3 L.s-1). The ignition of the kerosene/air mixture is ensured by means of a spark plug fixed on the upper part of the cone located at the outlet of the burner (see Figure 2). During the experiments, the composite sample is placed at 100 mm from the outlet of the burner cone in a sample holder equipped with a deflector to avoid any flame bypass (see Figure 3). The flame temperature is measured by a rack of seven K-type thermocouples having diameter of 1/8 inch positioned at the same spot as the test sample (100 mm from the burner cone), 20 mm apart from each other, and 25.4 mm above the central axis of the cone burner.

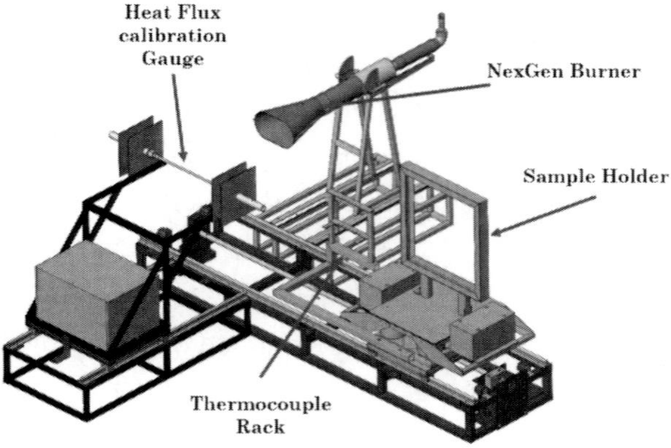

Figure 2. Overview of the NexGen Burner Test Bench.

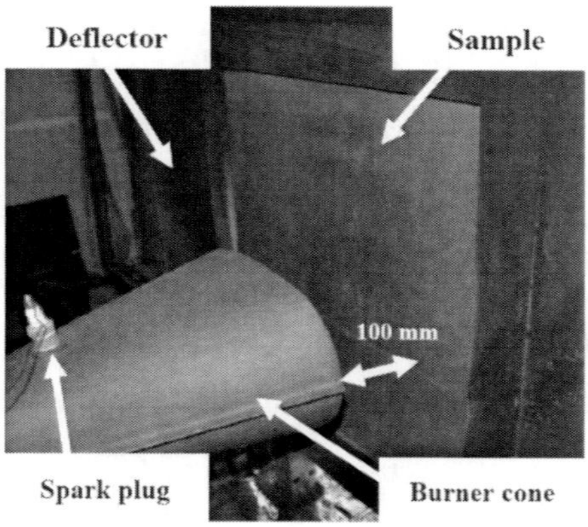

Figure 3. Overview of the sample fixation on the NexGen Burner Test Bench.

Studied Composite Materials

The two carbon-reinforced composites presented in this chapter have been manufactured and provided by the company DAHER. Among these, one is thermosetting i.e., carbon-phenolic and one is thermoplastic carbon-PEKK. The chemical representation of phenolic carbon is illustrated in Figure 4 and that of PEKK carbon is shown in Figure 5. The sample of thermoset considered in this chapter is a 6 plies Carbon-phenolic (1.5 mm in thickness) with a fiber mass fraction of 42%. This one will be compared to a 12 plies of thermoplastic carbon-PEKK (1.9 mm thickness) with a fiber mass fraction of 41%.

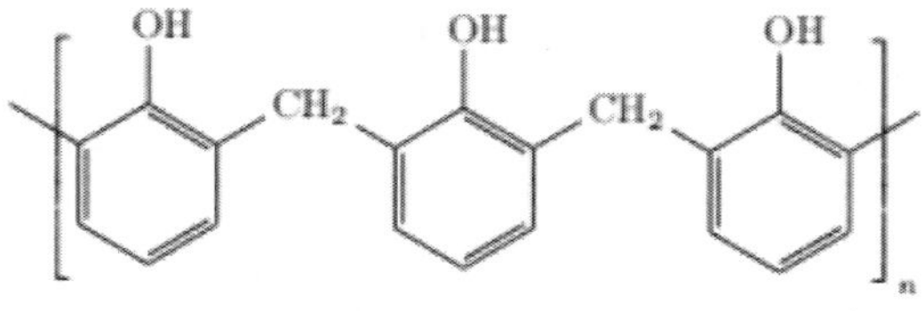

Figure 4. Chemical representation of phenolic carbon.

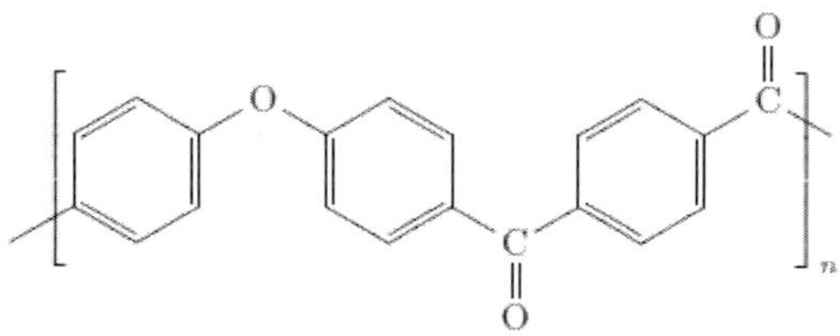

Figure 5. Chemical representation of carbon PEKK.

RESULTS AND DISCUSSION

Evaluation of the Pyrolysis Rate at Medium and Large Scale

To evaluate the pyrolysis temperatures and thermogravimetry (TG), measurements were performed with a SETARAM Setsys-16/18 TG under inert (argon) conditions and at atmospheric pressure. This device consists of a vertical-type TG in which the Al2O3 crucible is suspended from the electronically controlled balance [20]. All the experiments were carried out with an argon flow rate of 16.6 ml min^{-1}. A non-isothermal heating rate of 15 °C min^{-1} and a final temperature of 1000 °C were fixed, followed by an isothermal phase of 10 min at the final temperature before the cooling step. For phenolic and PEKK composites, sample masses of 7–9 mg directly cut from a virgin plate (1.5 mm thickness) were used, in order to limit the temperature gradient inside the samples.

In Figure 6 (a), it is presented the mass loss curves of carbon-phenolic and carbon-PEKK for a heating rate of 15°C/min. Using the evolution of the curves in the Figure 1, the materials can be characterized by three different behaviors (230°, 590°C and 780°C). It is observed that the both materials have a mass release about of 20% and the degradation beginning at approximately 200°C and decreasing slowly up to 900°C. However, the degradation of the carbon-PEKK start really around 515°C in comparison with phenolic Carbon which starts at 230°C highlighting a better thermal property of the carbon-PEKK. Moreover, thanks to the mesures performed with the DSC, it is shown in the Figure 6 (b) that, the degradation of the carbon PEKK need an energy more important in comparison with the phenolic Carbon. This observation comes justify the previous analysis made in the Figure 6 (a).

Now considering Figure 7 (a) and (b), it can be seen that in the ignition time, it is observed that the inflammation at the phenolic carbon level is greater than that of the PEKK carbon. This difference can be explained by the fact that the phenolic carbon losing mass more quickly (see Figure 6) creates a more flammable medium than that of PEKK carbon. This observation will also be found in the gaseous emissions part, as we will see that the phenolic carbon produces more pyrolysis gas. In addition, it is observed at the end of the test that the PEKK carbon exhibits a greater swelling than that of the phenolic carbon. Thanks to this swelling or delamination, it can therefore be justified that the temperature level of the rear face is lower than that of phenolic carbon. Because with this swelling, an insulating space has been created, thus contributing to a slowing down of the transmission of temperature by conduction.

This observation is made on a medium scale. It can also be found on a large scale, thanks to tests carried out on the NexGen burner. In fact, it can be seen in Figure 8 that at the end of the test, the phenolic carbon loses more plaice (3 plies) in comparison with the PEKK corbone (2 plies). As the plaice can be associated with the mass of the sample, it is deduced therefrom that the phenolic carbon loses mass more easily.

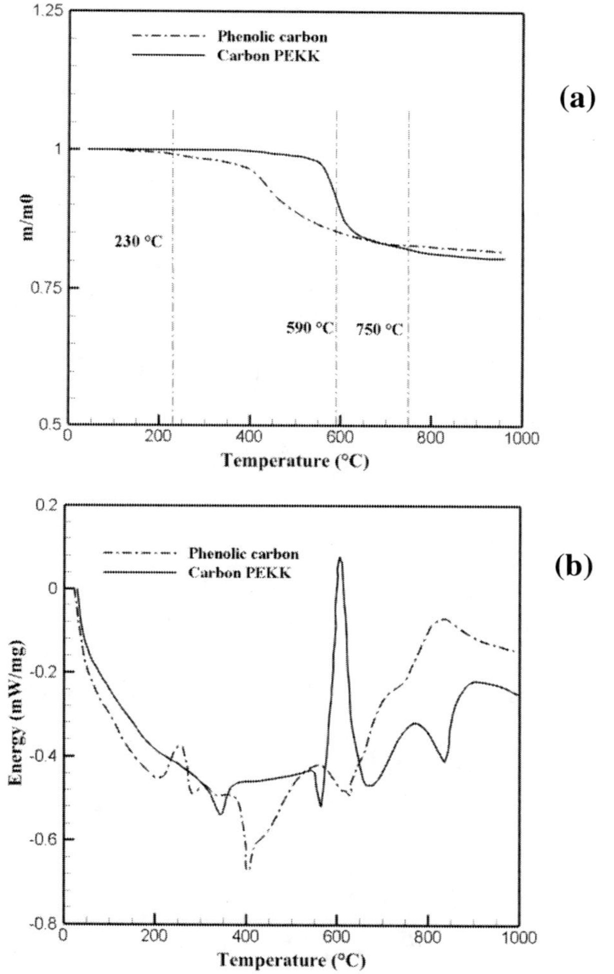

Figure 6. (a) TG and (b) DSC curves for Carbon

Continuing from the observations made previously, it can be seen in Figure 9 (a) and (b) that the mass loss of phenolic carbon is greater than that of carbon PEKK. This observation is similar between the medium scale (a) and the full scale (b). Thanks to this Figure 9, it is highlighted the fact that carbon PEKK has a better thermal performance in comparison with phenolic carbon. In addition, there is a similarity to the phenomenological level between the two scales, which justifies the fact that carrying out tests on medium scale provides a first approximation on the fire behavior of composite materials.

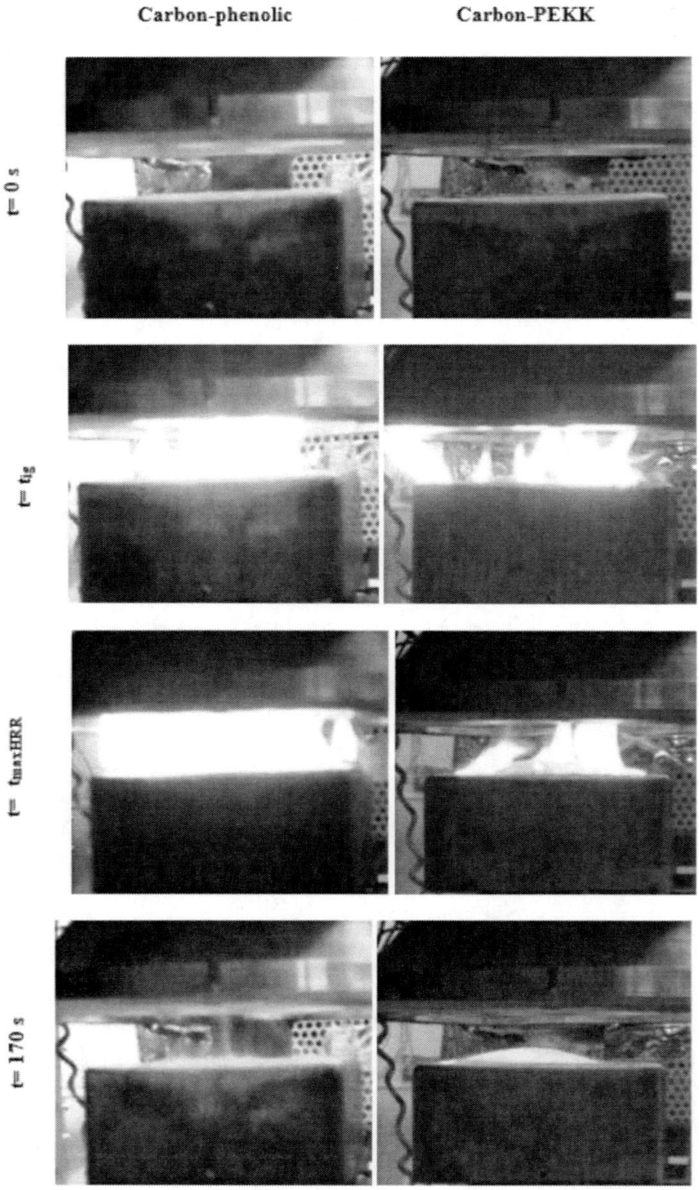

Figure 7. View of the main events during cone calorimeter exposition for both composites materials.

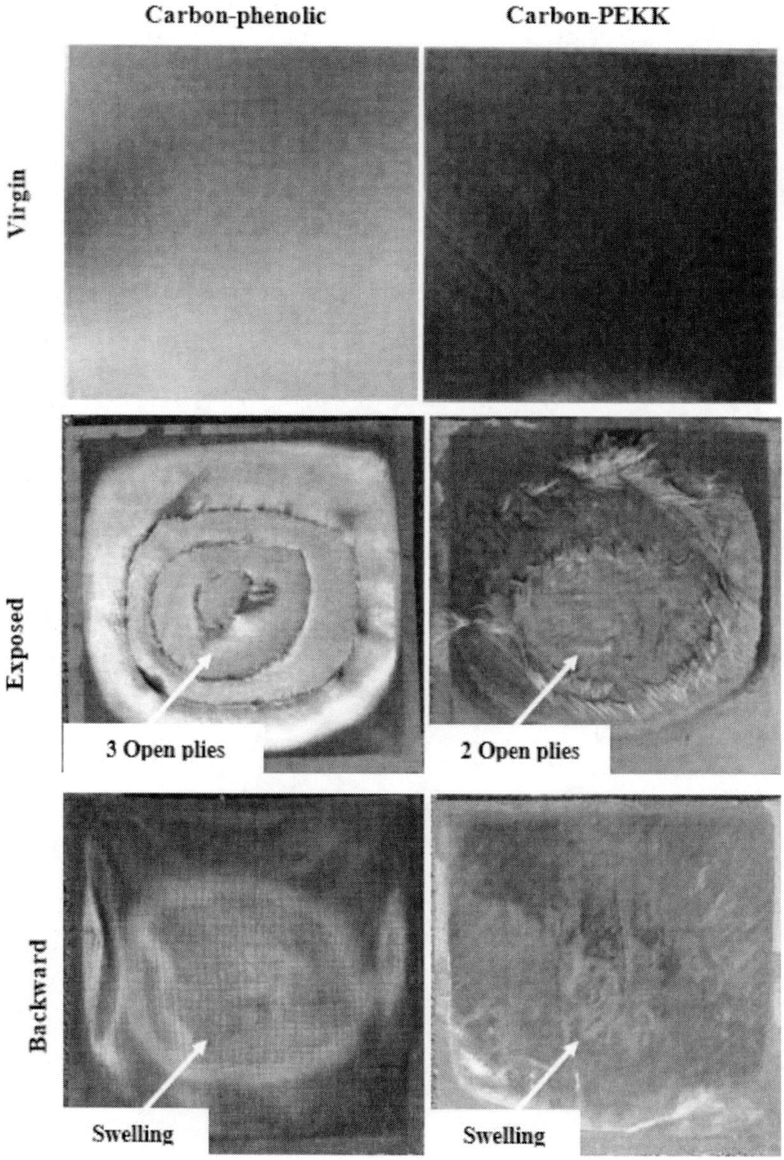

Figure 8. View of both composite materials before and after NexGen burner experiments.

Nevertheless, there is a great disparity between the results on the medium scale and those obtained on the full scale. This difference would come from the ratio of radiated flux and sample size.

At medium scale, for the carbon-PEKK, the mass loss demonstrates similar evolution. The mass is constant first up to 30 s, corresponding approximately to the ignition time, then a monotonous mass loss is observed until 160 s and then it starts to stabilize.

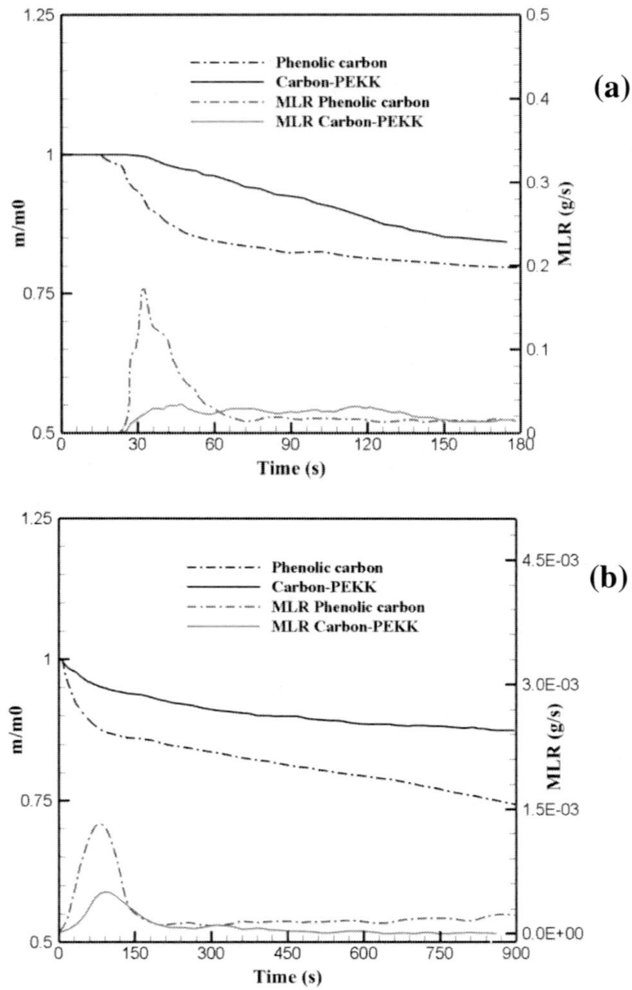

Figure 9. Pyrolysis rate: (a) at medium scale and (b) at large scale.

The obtained final mass loss equals 15% of the virgin sample mass. For the carbon-phenolic sample, different phases observed are visible. Initially, for almost 20 s (approximately the ignition time) the mass loss slightly changes, then an important and rapid mass loss between 20 s and 50 s is visible, and finally the mass loss decreases after 50 s until 23% of the virgin sample mass was consumed.

The differences observed in the mass loss behavior between the two materials are due to the nature of their resins. Moreover, the good thermal stability and the high melting point [20, 31] of the PEKK resin characterize its decomposition by a high yield of stable carbonaceous char (approximately 60% of the original resin mass at small scale [20]) which bring a superior flammability resistance [32]. The thermoset phenolic decomposition involves an important mass loss due to scission reactions between units along the polymer chain [29]. These reactions are respectively around 300°C and 400°C (corresponding to the second mass loss stage in Figure 6 (a) and (b)). The third stage of thermoset phenolic resin mass loss corresponds to the fusion of aromatic rings into carbonaceous char inducing superior fire performances and resulting in less mass loss rate than at the second stage [33].

Thermal Characterization of Compoiste Materials At medium and Large Scale

At medium scale, the average temperature profiles measured on the back faces are presented in Figure 10 (a). The values correspond to the mean temperature measured, using infrared thermography, on a 40 x 40 mm zone in the center of the backward face of the samples.

For the three materials the temperature evolution can be divided into two phases, a transient one with a quick increase of the BFT (Backward Face Temperature) followed by a constant growth of the temperature.

For the carbon PEKK, two phases are visible in the temperature curves. The first phase presents a fast increase of the temperature between 0 and 30 s followed by a short stabilization (at about 300°C) before a second temperature increase between 30 s and 90 s (of approximatively 100°C), Whereas Between 90 s and 180 s a final growth of approximately 100°C is observable. The latter can be associated with the stable phase. Regarding the carbon-phenolic sample, within the first 30 s the temperature raises 450°C before increasing again over 650°C which is the saturation temperature of the used infrared camera.

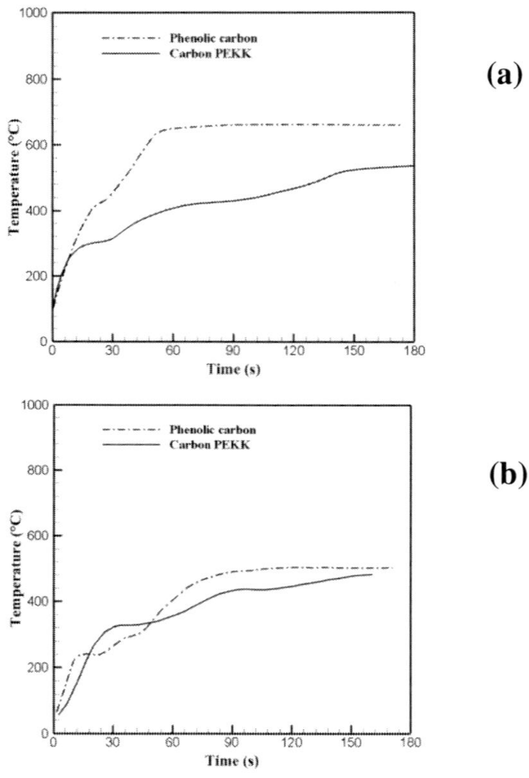

Figure. 10 Backface temperature: (a) at medium scale and (b) at large scale.

For the NexGen burner experiments, the average temperature evolution of a 1 cm x 1 cm area (i.e., a square of 4 pixels), located at the center of the rear face of the different samples, is computed for the totality of the test (15 minutes) and plotted on the Figure 10 (b). It is interesting to note that the overall change in temperature for both materials is similar. The temperature evolution can be split into two phases. A first phase corresponding to the rise in temperature, between 0 s and 150 s, presented on the Figure 10 (b), this phase might be associated with the significant degradation step visible on the MLR curve (Figure 6b). In second phase, the temperature remains almost constant (between 150 and 800 seconds). The slight change in temperature during this second phase agrees with the hypothesis put forward above, indicating that a thermal equilibrium is created in the material between the front face exposed to the flame and the rear face. At the same time, almost constant temperature leads to the very slow degradation of the composite observed on the mass loss curve. Contrary to the cone calorimeter experiments, the maximum temperatures obtained on the rear face of the materials are approximately equivalent, with temperatures between 500°C and 550°C. The heat beeing link to the mass loss rate, it is found the previous observations for the heat release rate (cf. Figure 11 (a) and (b).

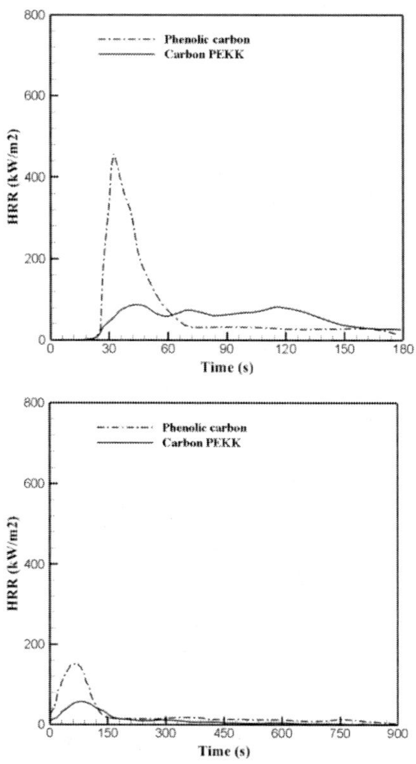

(a)

(b)

Figure 11. Heat Release Rate: (a) at medium scale and (b) at large scale.

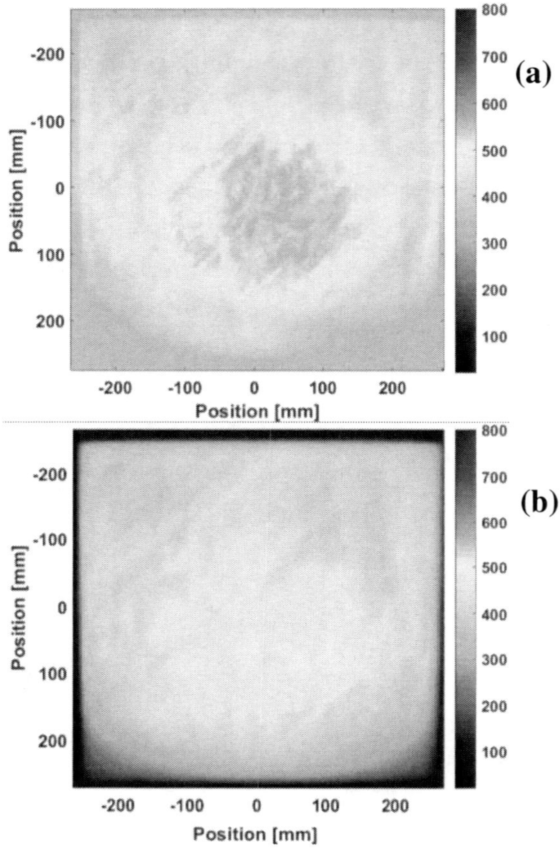

Figure 12. Backface Field temperature: (a) Phenolic carbon and (b) Carbon PEKK.

Regarding large scale experiments, the temperature fields were measured using the infrared camera placed approximately at 2.5 m from the back of the samples. Figure 12 presents instantaneous temperature field of the backward face of the carbon-PEKK sample (a) and the phenolic carbon sample (b) after about 150 seconds of exposure to the flame. By comparing these photographs of the back of the materials, it is possible to notice that the hottest areas visible in the figure correspond roughly to the most degraded areas visible in Figure 12. It is in these areas that the carbon fibers begin to oxidize leading to opening in the first plies on the carbon-PEKK and carbon-phenolic samples.

Characterization of Gas Emissions at Medium-Scale

The pyrolysis products observed on the Figure 13 for the two materials at four different temperatures were identified by mass spectrometry. The pyrolysis compounds identified, characterized by a high ignition potential, were then grouped in 5 categories (cf. Figure 13). The first category, corresponding to hydrocarbons, mainly contains substituted aromatic compounds such as benzene, toluene, and styrene. The second category corresponds to phenolic compounds which can be found in the two composite materials. Furans and ketones constitute respectively the third and the fourth families. The relative percentages of pyrolysis products associated with each category are plotted in Figure 13 for the different temperatures and materials.

Looking in detail at the pyrolysis products for the carbon-PEKK, one can see that the amounts detected were too insignificant to identify pyrolysis compounds below 450°C (Figure 13). This is the temperature where local instabilities at chain ends and chain branches might lead to the beginning of the thermal decomposition of the resin as observed on the mass loss curve on Figure 6 (a). At 590°C, once the thermal degradation has started, three hydrocarbons were mainly identified: butane, benzene and toluene (Figure 13). They might be released by the cleavage of the weakest bond in the PEKK monomer, which corresponds to the ether bond. At higher temperature (750 °C) a substantial release of phenol is observed (46%), probably associated with the cleavage of the ether bond, producing phenolic radicals which lead to phenol after hydrogen loss. Low amounts of aromatic compounds and ketones such as pentanone and p-benzoquinone are also suspected to be released by the cleavage of the ketone bond at this temperature (7%).

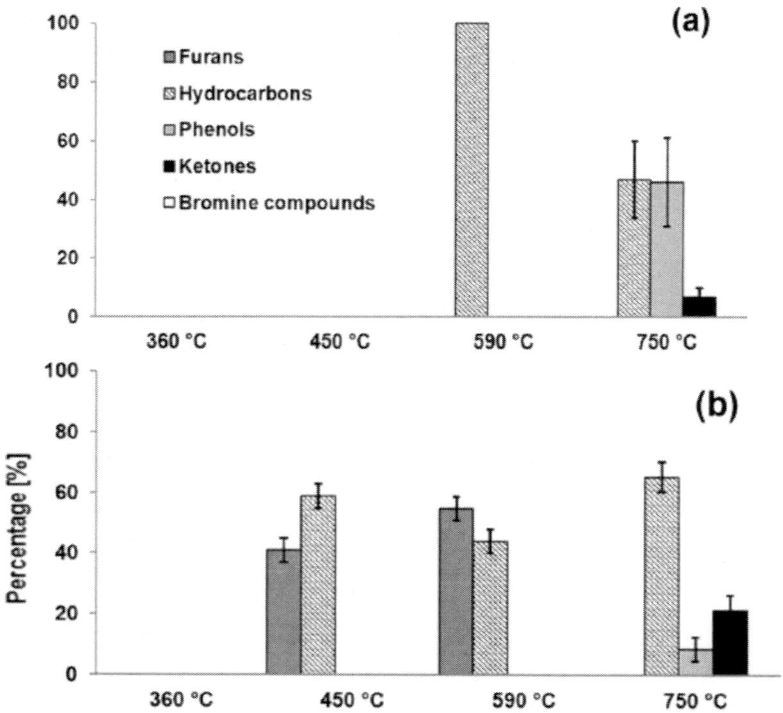

Figure 13. Gas emissions: (a) Carbon PEKK and (b) Phenolic carbon

The decomposition of the carbon-phenolic resin is known to begin at about 300°C with the dehydration of the resin, explaining why no pyrolysis product was detected at 360°C. At 450°C, the TG curve demonstrates that the reaction rate reaches a maximum, involving the release of butene (59%) and pentanone (41%). The latter might be formed by the combination of pyrolysis fragments with formaldehyde compounds either trapped during the resin curing or formed with the pyrolysis of phenol compounds as suggested by Lum et al. [39]. At 590°C, butene (39%) and pentanone (55%) appear to be the major pyrolysis products. The scission of a terminal benzyl bond at this temperature results in the emission of benzene as noticed by Trick et al. [40]. The pyrograms (Figure 13) also highlight the presence of toluene and xylene at this temperature in minor proportions (<10%). The latter might come from a terminal benzyl bond scission. Finally, at 750°C, the major pyrolysis products appear to be toluene (18%) and different isomers of dimethylbenzene (22%). Phenol and cresol (20%) are also detected as they are the major constituents of the resin.

The data of the combustibility and the ignitability of the two materials studied were obtained thanks to the TG and Py-GC–MS measurements described above and are given in Table 1. The various results demonstrate good thermal stability of the carbon-PEKK (cf. table 1). The latter thus has interesting characteristics as a means of fire protection for airplanes. Indeed, The carbon-PEKK appears to be the most suitable material as it has the highest degradation temperature and the lowest gas pyrolysis emission. It has similar performances to those of the carbon-phenolic resin in terms of mass loss and to those of the carbon-epoxy in terms of LFL.

Table 1. Fire characteristics of composite materials: carbon-phénolique and carbon-PEKK

		Carbon-PEKK	Phenolic carbon
Full scale	Open plies on the face exposed to the flame	2	3
	Maximum backface temperature (°C)	490	650
Medium scale	Mass loss (%)	19.6	18.4
	Ignition time (s)	25	20
	Maximum HRR Time (s)	40	38
	Pyrolysis temperature (°C)	590	230
	Gas emissions (%	17.6	21.9
	LFL (%)	0.67	0.44

CONCLUSION

In this chapter, the experimental characterization of the two materials on the two scales made it possible to highlight a few thermophysical properties of the latter, justifying that the PEKK composite material was more thermally stable than phenolic carbon. Indeed, the mass loss behavior obtained at medium and large scale demonstrates that the carbon-PEKK composite presents a superior thermal stability in comparison with the phenolic carbon. Moreover, this material presents a limited HRR value up to five times less than that of carbon-phenolic, thus limiting the fire spread. In the same manner, for both scales the temperature measures performed on the backside face demonstrate a similar behavior. Indeed, it is shown that the back-face temperatures for PEKK carbon are smaller than the one of carbon phenolic. This difference can be explain by the high swelling and delamination of the composite plies creating an insulating space during the fire test. These experimental results for the two materials demonstrate that both experimental scales give similar results despite the difference in the sample scale and thermal exposition (radiative vs flame). Moreover, gaseous emissions of the two carbon-based composite materials were evaluated as well as their lower flammability limits for four different pyrolysis temperatures. In this chapter, the goal was to present the thermal degradation of the materials used in this study in order to propose a fire classification based on their combustibility and ignitability. Combustibility was evaluated thanks to the degradation temperature and mass loss whereas ignitability considered pyrolysis gases with a high ignition potential and their lower flammability limits. To conclude, the classification of the two materials demonstrates that the carbon-PEKK presents the best fire performances. This composite is characterized by a high degradation temperature and a low mass loss. Moreover, it emits the lowest amount of pyrolysis products.

In addition, thanks to the similarity of the phenomena observed between the two scales, it is possible to say that in a first approximation perform experiment at medium scale provide a precise overview of the fire behavior of a material before performing large scale.

ACKNOWLEDGMENTS

The authors would like to thank the DAHER industry partner for its involvement in the project.

REFERENCES

[1] Marsh, G. (2011). Bombardier throws down the gauntlet with CSeries airliner. *Reinforced Plastics*, 55(6), 22-26.

[2] Shi, Y., Swait, T., & Soutis, C. (2012). Modelling damage evolution in composite laminates subjected to low velocity impact. *Composite Structures*, 94(9), 2902-2913.

[3] Kim, J. S., Jeong, J. C., & Lee, S. J. (2007). Numerical and experimental studies on the deformational behavior a composite train carbody of the Korean tilting train. *Composite Structures*, 81(2), 168-175.

[4] Marsh, G. (2007). Airbus takes on Boeing with reinforced plastic A350 XWB. *Reinforced plastics*, 51(11), 26-29.

[5] Fuchs, E. R., Field, F. R., Roth, R., & Kirchain, R. E. (2008). Strategic materials selection in the automobile body: Economic opportunities for polymer composite design. *Composites science and technology*, 68(9), 1989-2002.

[6] De Volder, M. F., Tawfick, S. H., Baughman, R. H., & Hart, A. J. (2013). Carbon nanotubes: present and future commercial applications. *Science*, 339(6119), 535-539.

[7] Mouritz, A. P., & Gibson, A. G. (2007). *Fire properties of polymer composite materials* (Vol. 143). Springer Science & Business Media.

[8] Grange, N., Chetehouna, K., Gascoin, N., & Senave, S. (2016). Numerical investigation of the heat transfer in an aeronautical composite material under fire stress. *Fire safety journal*, 80, 56-63.

[9] The International Organization for Standardization. (1998). *Aircraft environmental test procedure for airborne equipment resistance to fire in designated fire zones*.

[10] Federal aviation administration. (1990). *Powerplant installation and propulsion system component fire protection test methods, standard and criteria*. US Department of transportation.

[11] Zhang, J., Wang, X., Zhang, F., & Horrocks, A. R. (2004). Estimation of heat release rate for polymer–filler composites by cone calorimetry. *Polymer testing*, 23(2), 225-230.

[12] Hshieh, F. Y., & Beeson, H. D. (1997). Flammability testing of flame-retarded epoxy composites and phenolic composites. *Fire and materials*, 21(1), 41-49.

[13] Genovese, A., & Shanks, R. A. (2008). Fire performance of poly (dimethyl siloxane) composites evaluated by cone calorimetry. *Composites Part A: applied science and manufacturing*, 39(2), 398-405.

[14] ISO 5660-1: 2002. (2002). *Reaction-to-Fire Tests—Heat Release, Smoke Production and Mass Loss Rate—Part 1: Heat Release Rate (Cone Calorimeter Method)*.

[15] Fateh, T., Rogaume, T., & Richard, F. (2014). Multi-scale modeling of the thermal decomposition of fire retardant plywood. *Fire safety journal*, 64, 36-47.

[16] Feih, S., Mathys, Z., Mathys, G., Gibson, A. G., Robinson, M., & Mouritz, A. P. (2008). Influence of water content on failure of phenolic composites in fire. *Polymer degradation and stability*, 93(2), 376-382.

[17] Grange, N., Tadini, P., Chetehouna, K., Gascoin, N., Reynaud, I., & Senave, S. (2018). Determination of thermophysical properties for carbon-reinforced polymer-based composites up to 1000°C. *Thermochimica Acta*, 659, 157-165.

[18] Sorathia, U., Rollhauser, C. M., & Hughes, W. A. (1992). Improved fire safety of composites for naval applications. *Fire and Materials*, 16(3), 119-125.

[19] Chetehouna, K., Grange, N., Gascoin, N., Lemée, L., Reynaud, I., & Senave, S. (2018). Release and flammability evaluation of pyrolysis gases from carbon-based composite materials undergoing fire conditions. *Journal of analytical and applied pyrolysis*, 134, 136-142.

[20] Tadini, P., Grange, N., Chetehouna, K., Gascoin, N., Senave, S., & Reynaud, I. (2017). Thermal degradation analysis of innovative PEKK-based carbon composites for high-temperature aeronautical components. *Aerospace Science and technology*, 65, 106-116.

[21] Fink, J. K. (2017). *Reactive polymers: fundamentals and applications: a concise guide to industrial polymers*. William Andrew.

[22] Liu, Y. L., & Chen, Y. J. (2004). Novel thermosetting resins based on 4-(N-maleimidophenyl) glycidylether: II. *Bismaleimides and polybismaleimides. Polymer*, 45(6), 1797-1804.

[23] Rivière, L., Caussé, N., Lonjon, A., Dantras, É., & Lacabanne, C. (2016). Specific heat capacity and thermal conductivity of PEEK/Ag nanoparticles composites determined by Modulated-Temperature Differential Scanning Calorimetry. *Polymer Degradation and Stability*, 127, 98-104.

[24] Oztekin, E. S., Crowley, S. B., Lyon, R. E., Stoliarov, S. I., Patel, P., & Hull, T. R. (2012). Sources of variability in fire test data: a case study on poly (aryl ether ether ketone) (PEEK). *Combustion and flame*, 159(4), 1720-1731.

[25] Babrauskas, V. (1984). Development of the cone calorimeter—a bench-scale heat release rate apparatus based on oxygen consumption. *Fire and Materials*, 8(2), 81-95.

[26] Schuhler, E., Coppalle, A., Vieille, B., Yon, J., & Carpier, Y. (2018). Behaviour of aeronautical polymer composite to flame: A comparative study of thermoset-and thermoplastic-based laminate. *Polymer Degradation and Stability*, 152, 105-115.

[27] Benoit, V., L. Cédric, C. Alexis. (2014). Post fire behavior of carbon fibers Polyphenylene Sulfide-and epoxy-based laminates for aeronautical applications: a comparative study. *Materials & Design*, 63, 56-68.

[28] Maaroufi, M. A., Carpier, Y., Vieille, B., Gilles, L., Coppalle, A., & Barbe, F. (2017). Post-fire compressive behaviour of carbon fibers woven-ply Polyphenylene Sulfide laminates for aeronautical applications. *Composites Part B: Engineering*, 119, 101-113.

[29] Knop, A., & Pilato, L. A. (2013). *Phenolic resins: chemistry, applications and performance*. Springer Science & Business Media.

[30] Jiang, H., Wang, J., Wu, S., Yuan, Z., Hu, Z., Wu, R., & Liu, Q. (2012). The pyrolysis mechanism of phenol formaldehyde resin. *Polymer degradation and stability*, 97(8), 1527-1533.

[31] Patel, P., Hull, T. R., McCabe, R. W., Flath, D., Grasmeder, J., & Percy, M. (2010). Mechanism of thermal decomposition of poly (ether

ether ketone) (PEEK) from a review of decomposition studies. *Polymer Degradation and Stability*, 95(5), 709-718.

[32] Lochte, H. W., Strauss, E. L., & Conley, R. T. (1965). The thermo-oxidative degradation of phenol–formaldehyde polycondensates: Thermogravimetric and elemental composition studies of char formation. *Journal of Applied Polymer Science*, 9(8), 2799-2810.

[33] Trick, K. A., Saliba, T. E., & Sandhu, S. S. (1997). A kinetic model of the pyrolysis of phenolic resin in a carbon/phenolic composite. *Carbon*, 35(3), 393-401.

[34] Biasi, V. (2014). *Thermal modeling of the degradation of a composite material subjected to fire* (Doctoral dissertation).

[35] Walters, R. N., Hackett, S. M., & Lyon, R. E. (2000). Heats of combustion of high temperature polymers. *Fire and materials*, 24(5), 245-252.

[36] Trick, K. A., & Saliba, T. E. (1995). Mechanisms of the pyrolysis of phenolic resin in a carbon/phenolic composite. *Carbon*, 33(11), 1509-1515.

[37] Patel, P., Stec, A. A., Hull, T. R., Naffakh, M., Diez-Pascual, A. M., Ellis, G., & Lyon, R. E. (2012). Flammability properties of PEEK and carbon nanotube composites. *Polymer degradation and stability*, 97(12), 2492-2502.

[38] Macaione, D. P. (1992). *Flammability Characteristics of Fiber-Reinforced Epoxy Composites for Combat Vehicle Applications* (No. MTL-TR-92-58). U.S Army Materials Technology Laboratory.

[39] Lum, R., Wilkins, C. W., Robbins, M., Lyons, A. M., & Jones, R. P. (1983). Thermal analysis of graphite and carbon-phenolic composites by pyrolysis-mass spectrometry. *Carbon*, 21(2), 111-116.

[40] Trick, K. A., & Saliba, T. E. (1995). Mechanisms of the pyrolysis of phenolic resin in a carbon/phenolic composite. *Carbon*, 33(11), 1509-1515.

In: Critical Issues and Analysis … ISBN: 978-1-53618-738-0
Editors: A. Cablé and B. Manescau

Chapter 4

Sustainable Approach for Designing the Synthesis of Polysaccharide Based Bionanocomposites Fire Retardants

Dharmendra Kumar Jena
and Prafulla Kumar Sahoo*
Department of Chemistry, Utkal University,
Vani Vihar, Bhubaneswar, Odisha, India

Abstract

Fire, as one among the important tool throughout the human history, side by side, can bring severe disasters if out of control. In the world of materials the proper use of fire protection and prevention technology can

---

* Corresponding Author's Email: psahoochemuu@gmail.com.

significantly diminish fire hazards and contribute to save resources and lives. The present work is focusing to convert a low cost renewable polysaccharide to a high performance fire-retardant biomaterial by modification via grafting. Polysaccharides earth-abundant biomacromolecules, have gained increasing attention for their interesting properties and potential applications in the synthesis of polymer nanocomposites. Moreover, materials based on polysaccharides have wide range of applicability, predominantly due to realization of finite petroleum resources, sustainable and environmental concerns. This work includes a novel sustainable approach, highlighting the formulation of chicken eggshell powder (EP) modified starch-g-poly(N-isopropylacrylamide) (starch-g-PNIPAAm) bionanocomposites (BNCs) by emulsifier-free emulsion polymerisation and aiming to study the effect of EP on the properties of modified starch BNCs. Young's modulus and tensile strength of BNCs are found to be improved dramatically. The enhanced char forming ability of EP improves the thermal stability of BNCs at high temperature as investigated by thermogravimetric analyses carried out in an inert atmosphere. The cone calorimeter test revealed that the 4% w/v EP-based BNC resulted suppression on fire hazards in terms of reduction in PHRR (decreased by 33.3%), PSPR (decreased by 75.3%) compared with those of the control starch-g-PNIPAAm and could be attributed to the insulating barrier effect of EP. In addition, the fire retardancy of the BNCs is investigated from limiting oxygen index (LOI) test. The surface morphology and the elemental content of the collected char residues of BNCs after fire retardant test is analysed by FESEM and EDX. Further the oxygen barrier property of BNC with 4% w/v EP is reduced by 73% compared to starch-g-PNIPAAm. FTIR, XRD and TEM studies have reflected the resulting nanostructures and molecular interactions in the BNCs.

Keywords: polysaccharides, starch-g-PNIPAAm, fire retardancy, oxygen barrier property, bionanocomposites

INTRODUCTION

Designing of petroleum based polymer materials due to their lightweight, functionality and ease of processing has emerged both academic as well as industrial interest. Instead of conventional materials, materials derived from polymer find applications in aerospace, spacecraft, packaging, automobiles, sporting goods, electronics, civil construction, oil and gas industries etc. [1, 2]. Polymer materials are a major part of most of the components utilised by human being are well known for their high flammability and turn dangerous towards societal safety [3]. The inherent flammability property can extensively reduce their applications to large extent. Furthermore, polymer materials constitute of carbon-hydrogen backbones, which burn rapidly and a series of oxidation reactions (exothermic) arise with emission of toxic gases, corrosive smoke and heat. Combustion process involves two factors, such as combustible (reducing agent) and a combustive (oxidizing agent). Generally the combustive is oxygen present in air. The entire process begins with increase in temperature of polymeric materials because of heat source, to such extent where the temperature of materials reaches to their pyrolysis temperature. At this temperature polymer bond scissions are induced. The volatile fractions of resulting polymer fragments diffuse into air, creating combustible gaseous mixture (called fuel). These gaseous mixture are liberating heat when auto-ignition temperature (temperature where activation energy of combustion is attained) is reached upon ignition. The life span for combustion basically depends on the amount of heat generated during fuel combustion. When the heat liberated reaches to a certain level, different decomposition reactions are started in solid phase thereby producing more combustibles. Thus combustion cycle is maintained. Thus the un-controlled polymer combustion could result in severe damage and getting out of control [4, 5].

In USA nearly 4000 people and in Europe nearly 5000 people die every year because of fire accidents and about 0.3% of net domestic product loss occurred due to fire hazards [6]. In Great Britain 23100 automobile fire accidents are estimated by fire and rescue services [Crow Hurst 2014, Fire statistics@communities.gsi.gov.uk]. A report by development of US Fire analysis and research stated around 1240000 fire fatalities in US [7]. Between years 2001-2010, a Geneva based organisation namely Aircraft Crashes Record Office (ACRO) estimated 1662 aircraft accidents causing death of around 15962 people. After analysing the causes of these accidents, it was observed that, enhancement in fire retardancy of polymer materials can decrease the fire fatalities [8, 9]. A novel engineering to maximize fire safety is to avoid materials from igniting and reduce heat release rate once ignition occurs. This entire work primarily focuses on how fire retardancy by polymer materials can be accomplished and puts a serious demand for formulating fire retardants. The substance, on addition with polymer matrix provides necessary fire safety level and responsible for hindrance or retardance of fire, is popularly known as fire retardant. The goal of fire retardants is to minimize heat released below self-sustaining level of fire. The mechanistic activities involved in enhancing fire retardancy can be:

i. Modifying degradation mechanism of polymer materials.
ii. Directing combustion and decomposition reactions for releasing non-combustible species/heavy gases, which interfere during interchange of air and combustion gases.
iii. Generation of ceramic-like insulating char.
iv. Reduction of heat transference into polymer materials.
v. Formation of physical barriers for oxygen transmission.
vi. Physical integrity of polymer materials is well maintained.
vii. Enhancement in thermal conductivity or specific heat.

In addition, the enormous increase in the production and uses of petroleum based synthetic materials for packaging applications, automobiles, building materials etc. have put serious pressure not only on environment but also on economy due to their non-renewability, inferior biodegradability and high cost [10, 11]. According to survey, nearly 20% volume of municipal waste was generated in Germany and North America from synthetic polymeric films (non-biodegradable) [12]. About 76.7 million tons of waste obtained from packaging materials has been reported by Municipal Solid Waste (MSW) report-2008 of U. S. Environmental Protection Agency (EPA). Further, about 13.2% of these packaging wastes were recycled, 9.8% were recovered while the rest were added in landfills [13]. Central Pollution Control Board's (CPCB) report 2015 observed that in India about 6.92% of MSW is from plastic waste. The data of plastic waste generation estimated about 25940 tons of waste are generated from plastics per day in India.

And only 60% or 15600 tons plastic wastes being recycled and rest are ends up in oceans, seas or lands without recycling. A report from plastic processing industry (2018) stated, from year 2017 to 2022 polymer consumption is likely to rise at 10.4%. Another report from Ministry of Petroleum and Natural gas, India predicts by the year 2022 annual per capita plastic consumption would be around 20 kgs. CPCB's report (2018) stated that the total plastic waste generation in India was 660787.85 Tons per annum (TPA).

The occurrence of non-biodegradable materials brings major problems like disposal problems, flash floods, white pollution and global warming [14]. The issues of economic and environmental impact associated with petroleum based synthetic polymeric materials have led to search for formulating materials which are both biodegradable and recyclable with significant fire retardancy [15, 16]. Recently biopolymer/polysaccharides have extensively been studied for designing biomaterials due to their wide availability, low density, renewability, biodegradability and cost competitiveness [17, 18, 19]. Moreover, natural biopolymers are superior compared to synthetic biodegradable polymers due to more biodegradation as well as renewability [20]. However the use of biopolymers in synthesizing materials has some key demerits such as water absorption, poor compatibility with hydrophobic polymer matrices and minimization of processing temperature [21, 22]. The adsorbed water in biopolymer readily reduces interfacial adhesion to polymer matrices, which induces some plasticizing effect, resulting inferior mechanical properties [23]. Moreover, research community have shown extensive attention towards developing polymer nanocomposites (NCs) with nano-sized clays/fillers incorporation. These inorganic particle acted as a reinforcement agent as well as provide superior physical and chemical properties to the synthesised novel materials compared to neat polymer [24]. It is the fact, why many industries are paying huge interest by investing billion dollars annually to develop polymer/clay NCs. It is surveyed nearly 76-80 per cent of polymer/clay NCs are utilised in packaging, aeronautic and automotive industries [25].

Samal et al. reported the preparation of Natural fiber composites by grafting methyl methacrylate (MMA) onto chemically modified rice straw

(RS) in an aqueous medium with and without sodium silicate (SS) additive [26]. The temperature of decomposition (TD) was found to be 205°C for RS-g-PMMA and 250°C for the (RS-g-PMMA)/SS composite. Residual mass for RS-g-PMMA and (RS-g-PMMA)/SS composite was found to be 9% and 15%, respectively indicating with SS addition, (RS-g-PMMA)/SS composite becomes somewhat resistant to thermal action. The heat release rate (HRR) value of RS, RS-g-PMMA and (RS-g-PMMA)/SS was found to be 554±12, 543±12 and 423±12 kW/m2 respectively. The enhancement in flammability properties of (RS-g-PMMA)/SS attributed to the formation of charred residue with silicate layer present in it, which acted as an insulating barrier by decreasing heat and mass transference between the flame and the samples. Chemically modified jute fiber (Tossa variety) grafted 2-ethyl hexylacrylate (JF-g-PEHA) fire retardant composites were successfully prepared with montmorillonite (MMT) incorporation and strong interactions were studied by Sahu et al. [27]. The HRR and mass loss rate (MLR) for JF-g-PEHA/MMT composite was 567.8 kW/m2 and 0.032 g/s was observably reduced with MMT incorporation in JF-g-PEHA matrix due to barrier effect. Mallick et al. developed fire retardant bionanocomposites (BNCs) by taking the advantages of starch polysaccharide [28]. The incorporation of mica into starch-g-poly butyl acrylate (starch-g-PBA) was studied, which supports the improved thermal stability, water absorbency and biodegradability. However, the enhanced fire retardancy of novel starch-g-PBA/mica BNC, supplying fire protection by delaying the ignition time with greatest extension of time at a peak HRR on exposure to fire. Jena et al. demonstrated the affinity between layered kaolin and cellulose through PBA, creates strong interfacial adhesion, thereby formulating cellulose-g-PBA/kaolin NCs [29]. The peak HRR value of cellulose-g-PBA copolymer was found to be 1112 kW/m2 and that for cellulose-g-PBA/20% w/v kaolin NCs was 978 kW/m2. Likewise total heat release (THR) for cellulose-g-PBA/20% w/v kaolin NCs was reduced to 56 mJ/m2 indicating improved fire prevention. This could be explained by the formation of char residue, which acted as a protective barrier by reducing the heat and mass transfer between the flame and underlying materials.

The present work is highlighting the exploration of a promising method (emulsifier-free emulsion polymerisation) for combining biopolymer with synthetic polymer, simultaneously introducing clay/filler particles in polymer matrices to design a novel breed materials. Further, this work is addressing the various aspects for polymer combustion.

Within the broad family of polysaccharides, starch is such a non-toxic, inexpensive abundantly available most promising candidate for synthesising biodegradable materials due to its applicability and feasibility [30-34]. Starch is composed of linear amylose (poly-α-1,4-D-glucan) and branched amylopectin (poly-α-1,4-D-glucan and poly-α-1,6-D-glucan) [35, 36]. However, the limitations of starch-based materials including inferior mechanical properties [37], fast degradation rate and poor water resistance [38], currently restrict their practical applications. Previously, van Krevelen et al. [39] investigated that plastic films have exhibited high elongation at break (200-800%) and tensile strength (10-80 MPa). In comparison, starch films are brittle and exhibit elongation ranging from 4 to 8% [40-42]. Thus the modifications of starch by casting [43], melt-mixing [44], blending [45], grafting [46], solution method or by incorporation of clay particles [47] to impart desired properties, such as mechanical, thermal, biodegradation and fire retardant properties, in particular, for starch-based biomaterials are of both interest and need.

Poly(N-isopropylacrylamide) (PNIPAAm) [48, 49] a well-known environmentally sensitive polymer has been extensively studied for preparing biopolymer-based functional materials to enlarge the potential applications of biopolymers [50, 51]. Wang et al. demonstrated the combination of bacterial cellulose (BC), 4,4'-methylenediphenyl diisocyanate (MDI)-modified BC, would greatly enhance the mechanical property of PNIPAAm composite hydrogel. And the MDI-modified BC/PNIPAAm composite hydrogel with an anisotropic layered structure exhibited compressive strength 40 times higher than virgin PNIPAAm gel due to the strong interaction between PNIPAAm and MDI-modified BC [52]. Chang et al. reported efficient way to create double network hydrogel consisting biopolymer and synthesized PNIPAAm with interpenetrating polymer network (IPN) strategy. In addition, the strong hydrogen-bonding interaction, homogeneous morphology and good compatibility between PNIPAAm and cellulose result in high mechanical strength as well as good temperature-sensitive properties in the hydrogel [53]. Tan et al. investigated the structure and properties of PNIPAAm/starch-based nanospheres thermo-responsive hydrogels exhibiting rapid response rate and good mechanical strength. The results attributed to the flexible chain lengths between the starch-based nanospheres as such; the stress may consistently be distributed between the flexible polymer chains and well defined network structure of the hydrogel [54].

More importantly, it has a great interest to seek for reinforcing bio-fillers in the polymeric matrix, which can provide superior performances, such as improved barrier properties, mechanical properties, fire retardancy and thermal stability leading to the future generation of biomaterials [55-58]. Chicken eggshell an environmentally friendly, abundant, inexpensive, renewable and low density bio-filler has generated attention for creating new value in polymer nanocomposites [59]. It is an industrial byproduct constitutes by a three- layered structure, an inner lamellar (or mammillary) layer, a spongy (calcareous) layer and the cuticle on the outer surface and contains about 95% calcium carbonate (CaCO3) in the form of calcite and 5% organic materials such as type X-collagen, sulfated polysaccharides, and other proteins [60]. It has significant uses for various applications such as packaging [61], nucleating agent [62], soil conditioner [63], green and economical adsorbent [64] and intumescent flame retardant coatings [65]. It has another most useful application as a filler for synthetic polymeric materials including polyethylene [66], epoxy [67], polypropylene [68]. However, there have been few works reported on starch-based materials reinforced by chicken eggshell [30, 56, 62]. Jiang et al. reported that the incorporation of 2 wt% of chicken eggshell powder increased the tensile strength 4.69 MPa for the corn starch-eggshell powder composite films. It could be attributed to strong molecular interaction and biocompatibility between these two components [56]. Hence, the use of chicken eggshell based CaCO3 as a bio-filler will reduce the reliance on non-renewable commercial CaCO3 and opens a path for its recycling in a cost-effective manner.

On the basis of these discussions, it is quite reasonable to expect the synthesis of chicken eggshell powder (EP) modified starch-g-PNIPAAm bionanocomposites (BNCs) by cost-effective emulsifier-free emulsion polymerisation with enhanced physical properties. Emulsifier-free emulsion polymerization method rules out any chances of removal of residual added surfactants or emulsifiers from the polymeric matrixes whose removal indeed, is a complicated, time-consuming and costly method. Again in a conventional emulsion method the presence of absorbed surfactants leads to somewhat variable properties and applications. In view of this, emulsifier-free emulsion polymerization method is chosen for the synthesis of BNCs and can be visualized as a viable alternative to the conventional emulsion method [69, 70, 71]. Previously guar gum (GG) based nanocomposites are prepared by the exfoliation of montmorillonite (MMT) onto the virgin matrix guar gum-g-polybutyl acrylate (GG-g-PBA). The cone calorimetry analysis revealed that for GG-g-PBA+10% MMT nanocomposite the peak HRR is reduced by 32.17%. GG-g-PBA+10% MMT nanocomposite demonstrated highest Young's modulus of 8.3±0.2GPa and tensile strength of 92±5MPa GG-g-PBA copolymer had a Young's modulus of 0.5±0.1GPa and tensile strength of 51±6MPa. This could be attributed to nano level interaction between clay and polymer matrix enabling greater possibility of stress transfer from the matrix to clay [69]. We also reported the synthesis of carboxymethyl cellulose-g-polyacrylonitrile/montmorillonite ((CMC-g-PAN)/MMT) nanocomposites exhibiting dramatically reduction in peak HRR of 536±03 kW m−2 in comparison to neat matrix. As well as, the tensile strength of the nanocomposite was also increased by around 41% with 5% w/v MMT contents [70]. The main objective of this study is to investigate the effect of different concentrations of EP bio-filler on improving thermal, mechanical, oxygen barrier properties, fire retardancy and biodegradation of the novel modified starch BNCs for industrial applications.

MATERIALS AND METHODS

Materials

Monomer, N-isopropylacrylamide (NIPAAm, assay 97%) was provided by Sigma-Aldrich, USA. Starch was a product of Qualigens, Mumbai, India. Ammonium persulfate (APS, assay 98%, analytical grade), N, N'-methylene bisacrylamide (MBA, assay 99%, chemically pure) were supplied by Aldrich (analytical grade). N, N, N', N'-tetramethylethylenediamine (TEMED, assay 99%) was purchased from Fluka (Busch, Switzerland). Ferrous ammonium sulfate (assay 98%, analytical grade) was from E.Merck, India. The chicken eggshells, collected from waste of the university cafeteria were washed in deionised water thrice to remove any impurities and then dried in a hot air oven at 120°C for 2 h [58]. The dried chicken eggshells were crushed and grinded by a mechanical attritor. Finally, the chicken eggshells powder (EP) was oven dried at 120°C until a constant weight was obtained and stored in a desiccator prior to use.

Synthesis of Starch-g-PNIPAAm Copolymer and (Starch-g-PNIPAAm) /EP BNC

Table 1. Composition of the feed mixture and preparation conditions

Sample code	Starch (g)	NIPAAm (g)	APS (g)	MBA (g)	TEMED (ml)	EP (% w/v)	% grafting
S1	1	1.0	0.2	0.1	0.2	–	42.3±1
S2	1	1.1	0.2	0.1	0.2	–	49.2±1
S3	1	1.2	0.2	0.1	0.2	–	54.7±3
S4	1	1.3	0.2	0.1	0.2	–	61.3±1
S5	1	1.4	0.2	0.1	0.2	–	53.2±2
S6	1	1.5	0.2	0.1	0.2	–	49.7±2
BNC1	1	1.3	0.2	0.1	0.2	1	–
BNC2	1	1.3	0.2	0.1	0.2	2	–
BNC3	1	1.3	0.2	0.1	0.2	3	–
BNC4	1	1.3	0.2	0.1	0.2	4	–

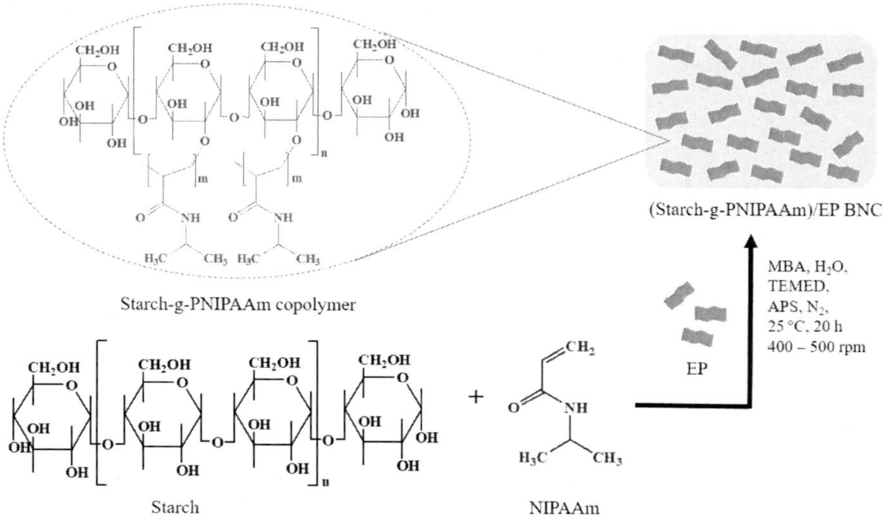

Figure 1. Schematic illustration for the synthesis of (starch-g-PNIPAAm)/EP BNC. [Copyright (2019), with permission from Elsevier [Details given in Reference No. 71]].

A requisite amount of starch (1.0 g) was added with 30 ml of deionized water in a 500 ml reactor equipped with a mechanical stirrer, which was stirred about 250 rpm for 15 min. The reaction temperature was controlled by placing the reactor in a thermostat water bath at 80°C. After homogenizing the mixture, the monomer NIPAAm and the crosslinker MBA were simultaneously added and again stirred for 15 min. Then the initiator APS and accelerator TEMED were added to the reaction mixture. The solution was stirred at 400–500 rpm under nitrogen atmosphere at 25°C for 20 h to synthesize the graft copolymer as shown in Table 1. Then (starch-g-PNIPAAm)/EP BNCs were prepared by taking the same quantity of NIPAAm and starch in the reaction vessels with 30 ml of deionized water followed by addition of bio-filler EP (1%, 2%, 3%, 4% w/v) to each vessel and stirred for 15 min. The grafting procedure described as above is followed by addition of MBA, TEMED and APS with continuous stirring under nitrogen atmosphere. Details about the synthesis of BNCs are listed in Table 1. Then the polymerisation was arrested by addition of a 0.1 mol.dm-3 ferrous ammonium sulfate solution. The coagulated products were purified by washing with deionized water thrice, oven dried at 40°C for 24 h and stored in a desiccator prior to characterization. The structure of the so synthesized BNC was illustrated in Figure 1 [58, 69].

Methods of Characterization

The FTIR spectra of starch, EP, starch-g-PNIPAAm, (starch-g-PNIPAAm)/EP were recorded on Perkin Elmer Paragon 500 FTIR spectrophotometer with KBr pellets. Spectrum in the optical range of 400-4000 cm-1 were obtained by averaging 16 scans at a resolution of 4 cm-1. Baseline was corrected for each spectrum. X-ray diffraction (XRD) patterns of the BNC and raw materials were studied using Rigaku X-ray machine, operating at 40 kV and 150 mA with Cu Kα (λ=1.54 Å), scanning from 5° to 40°.

The morphologies of BNCs and filler dispersion was visually investigated by means of transmission electron microscope (TEM) (H-7100, Hitachi Co.), operated at 100 kV.

Study of Properties

Thermal properties were measured by using a Shimadzu DTG 50 thermal analyser. It was carried out under N2 atmosphere from 30°C to 800°C at a heating rate of 20°C/min with a flowing rate of 100 mL min-1. The TGA experiments were performed in duplicate.

Tensile bars were obtained on a Van Dorn 55 HPS 2.8F mini injection molding machine under the following processing conditions: a melt temperature of 150°C, a mold temperature of 25°C, an injection speed of 40 mm/s, an injection pressure of 10 MPa, a holding time of 2 s, with a total cycle time of 30 s. Tensile measurements on injection molded samples of nanocomposites were performed according to ASTM D-638-00 using an Instron test machine Model 5567. Tests were carried out at a crosshead speed of 50 mm/min and a 1 kN load cell without the use of an extensiometer. All tests were performed at room temperature and the results were the average of five measurements. The highest value of standard deviation was 15%.

The fire retardancy of the prepared samples was assessed by limiting oxygen index (LOI) test according to ASTM D 2863 apparatus from Fire Instrument Research Equipment Ltd. with a digital read out of oxygen concentration to + 0.1%. The dimension of the specimens used for this test were of 100×6.5×3 mm3. Cone calorimetry was performed according to ISO 5660 standard at an incident flux of 50 kW m-2 of each specimens with dimensions 100×100×4 mm3. The results reported herein are the average of three tests for each sample. Char residue was examined with a SU3500 model FESEM. The imaging were carried out in liquid N2 and a few nanometer thick layer of gold was sputtered on to the sample surface prior to imaging. The relative elemental content of the char residue was analysed by energy-dispersive X-ray analysis (EDX).

Oxygen permeability of the samples was measured according to ASTM F 316-86 by using Oxygen Permeation Analyser (PMI instrument, model GP-201-A, NY). For testing oxygen permeability the samples were converted to films of thickness 5 mm. The results obtained are the average of three tests for each sample.

The biodegradation was studied by sludge water via weight loss. Microorganisms such as yeast, bacteria, fungi etc present in sludge water are mainly responsible for the biodegradation of waste materials. The sludge water was collected from tank areas receiving domestic wastewater and toilet, in a polypropylene container and brought to the laboratory immediately. After settling for about 1 h, the total solid concentration was increased to 5000 mg/l. The polymer sample (0.3 g) and activated sludge water were incubated together in a sterilized vessel at room temperature (28 ± 2°C) for 30, 60, 90, 120 and 150 days. Duplicate samples were removed at time intervals for biodegradation study via weight loss. Vessels containing polymer samples without sludge water were treated as controls.

RESULTS AND DISCUSSION

FTIR Spectrum Analysis

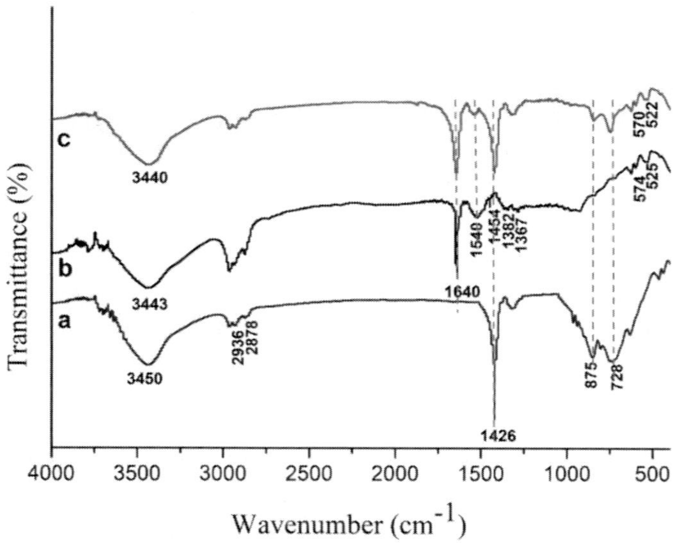

Figure 2. FTIR spectra of (a) EP, (b) starch-g-PNIPAAm copolymer and (c) (starch-g-PNIPAAm)/EP (4% w/v) BNC.
[Copyright (2019), with permission from Elsevier [Details given in Reference No. 71]].

The synthesis of (starch-g-PNIPAAm)/EP BNCs was confirmed by FTIR spectrum of bio-filler EP, copolymer starch-g-PNIPAAm and (starch-g-PNIPAAm)/EP BNC as well as the results are shown in Figure 2. FTIR spectrum of EP showed the characteristic peaks at 1426, 875, and 728 cm−1 are due to the asymmetric stretching vibration, out-of plane bend vibration and in-plane bend vibration for CO32− groups in CaCO3, respectively (Figure 1a) [56]. The peaks centered at 2936 and 2878 cm−1 are attributed to the asymmetric and symmetric stretching vibrations of C-H and -CH2- groups, respectively, revealing the presence of organic contents in EP bio-filler. Generally, organic contents present in EP resulting uniform dispersion of EP throughout starch-g-PNIPAAm virgin matrix because of strong affinity between matrix and bio-filler. The sharp peak at 3450 cm−1 is because of stretching vibration for structural water molecule [56, 57].

A characteristic absorption peak at 3443 cm -1 attributed to –OH stretching vibration of starch is observed in FTIR spectra of starch-g-PNIPAAm as shown in Figure 1b. The peaks at 1640 and 1540 cm-1 due to amide I and amide II of PNIPAAm respectively indicating successful grafting of PNIPAAm on starch [52]. The peak at 1454 cm-1 is appearing for the bending vibration of C-H, while the peaks at 1367 and 1382 cm-1 are appearing for the isopropyl group of PNIPAAm which provide evidence regarding grafting [49].

FTIR spectrum of (starch-g-PNIPAAm)/EP BNC showing intensity variation and shift of peak 3440 cm−1 due to –OH stretching vibration of starch is demonstrated in Figure 1c. The intensities of peaks referred to vibration of CO32− groups present in EP are decreased (Figure 2c) which are not appeared in Figure 1b. Also variations of intensities of peaks for amide I and amide II in the spectrum of BNC are observed. Further, the bands at 525 and 574 cm-1 attributed to pyranoid ring of starch shifted to 522 and 570 cm-1 implies the strong interaction between starch-g-PNIPAAm and EP for synthesising aimed BNC (Figure 2c) [56].

X-Ray Diffraction (XRD) Analysis

XRD patterns of (starch-g-PNIPAAm)/EP BNCs, starch-g-PNIPAAm copolymer and EP were undertaken to gather structural information about prepared materials (Figure 3). EP displays a characteristic peak at 2θ value 29.6°, corresponds to d-spacing of 3.01 Å (Figure 3) [30]. However, this diffraction peak is not present in S4 (Figure 3). For BNC1 a strong reflection peak centered at 28.7° is appeared. This reveals an increase in d-spacing to a higher value of 3.12 Å. This peak is appeared for all BNCs, and an increase in this peak intensity with EP addition was noticed (Figure 3). The increase in d-spacing confirmed that starch-g-PNIPAAm matrix was inserted within EP galleries leading to intercalation without disorderness [56].

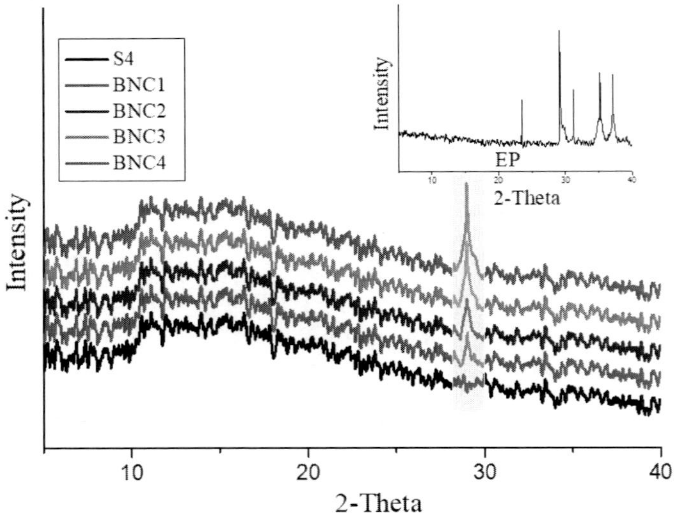

Figure 3. XRD patterns of EP, starch-g-PNIPAAm copolymer and (starch-g-PNIPAAm)/EP BNCs.
[Copyright (2019), with permission from Elsevier [Details given in Reference No. 71]].

TEM Analysis

TEM images of BNC are presented in Figure 4 corresponding to the inferences given in the XRD result. The uniform dispersion and intercalation of EP bio-filler throughout starch-g-PNIPAAm matrix were clearly noticed in Figure 4a. Further, the bright phase represents control starch-g-PNIPAAm matrix as well as dark phase represents bio-filler EP have been visualised in TEM image (Figure 4b) [67]. Moreover, these concludes a strong affinity between bio-filler and polymer matrix, and could resulting an excellent enhancement in mechanical properties of BNCs.

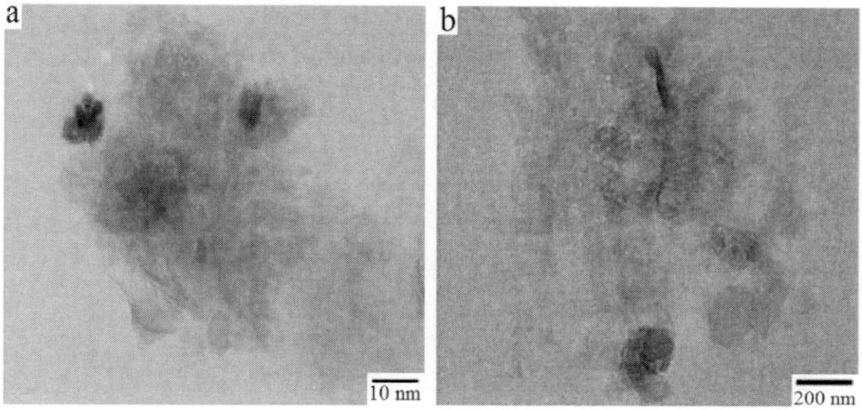

Figure 4. TEM images of (a) BNC4 at 10 nm and (b) BNC4 at 200 nm. [Copyright (2019), with permission from Elsevier [Details given in Reference No. 71]].

Thermogravimetry Analysis

TGA demonstrates necessary information regarding thermal degradation and the effect of EP on the thermal stability of prepared materials by the measurement of materials weight loss as the function of temperature. Figure 5 shows TGA curves for starch-g-PNIPAAm matrix S4 and BNCs with different EP contents in N2 atmosphere. At initial stage, the minor weight loss achieved before 100°C, is because of elimination of water and low molecular weight molecules [43]. Furthermore, the major decomposition of each prepared materials begin above 300°C, which is due to depolymerisation, elimination of hydroxyl groups and random carbon chains scission [44, 46, 49]. 50 wt% loss for virgin S4 occurred at temperature of 314.8°C, whereas a 50 wt% loss for BNC1, BNC2, BNC3 and BNC4 occurred at temperature about 319.8, 323.2, 327.9 and 330.9°C respectively. The char residuals left after weight loss at 800°C was about 3.1% for S4 as well as 5.6%, 8.6%, 11.8%, 14.8% for BNCs with 1%, 2%, 3%, 4% w/v EP contents respectively. The above discussions suggesting uniform dispersion of EP onto virgin matrix can remarkably enhance the thermal stability of bio-filler modified starch BNCs at higher temperature and promotes char formation [30, 69].

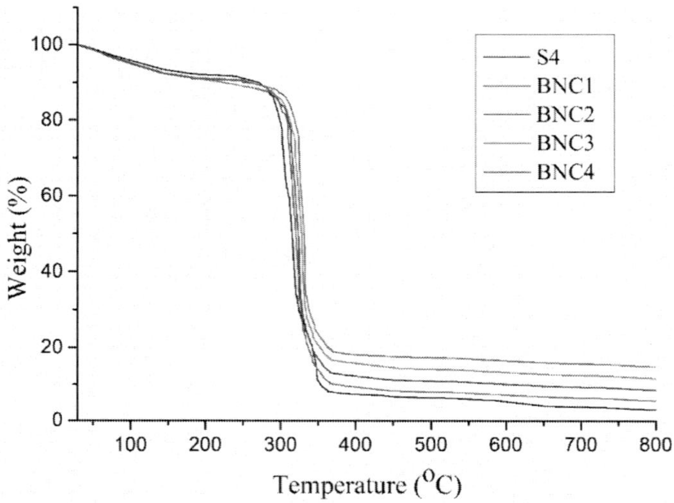

Figure 5. TGA thermograms of S4 and BNCs with different EP contents. [Copyright (2019), with permission from Elsevier [Details given in Reference No. 71]].

Mechanical Properties

Table 2. Comparative data of mechanical properties of starch-g-PNIPAAm copolymer and (starch-g-PNIPAAm)/EP BNCs

Sample code	YM (GPa)	TS (MPa)	E%
S4	0.3±0.1	24±1	80.3±3
BNC1	1.4±0.1	32±1	98.2±1
BNC2	2.3±0.2	41±3	106.3±1
BNC3	3.7±0.3	53±2	112.5±2
BNC4	4.6±0.1	67±1	129.4±6

YM, Young's modulus; TS, tensile strength; E%, elongation at the break.
[Copyright (2019), with permission from Elsevier [Details given in Reference No. 71]].

Evaluation of mechanical properties of materials including E%, TS and YM were assayed and results are summarized in Table 2. EP content had a significant effect on E%, YM, TS for BNCs. S4 had E% value of 80.3±3% and for BNC4, it was found to be 129.4±6%. TS increased from 24±1 MPa to 67±1 MPa and YM increased from 0.3±0.1 GPa to 4.6±0.1 GPa with an increase in EP content from 0 to 4% w/v. This could be explained because of strong interfacial adhesion between EP based CaCO3 and starch-g-PNIPAAm matrix, leading to decrease in mobility of matrix and effective stress transfer from control matrix to filler [30]. The existence of strong interaction was agrees well with the TEM and FTIR analysis. The increase in mechanical properties with EP loadings was also previously reported [55, 66, 67]. This could also be confirmed by XRD results (Figure 3), where the wide spacing of EP enables the penetration of starch-g-PNIPAAm matrix, was probably the reason for the enhanced mechanical properties of EP modified starch based BNCs [47].

Fire Retardancy

Cone calorimetry tests have been undertaken to investigate the influences of bio-filler EP on the fire retardancy of neat starch-g-PNIPAAm matrix, simultaneously to simulate the fire scenario. Key parameters like TTI, PHRR, THR, MLR and PSPR are listed in Table 3 while Figure 6 provides cone curves for some parameters observed at the time of cone calorimetry tests.

Table 3. Cone calorimetry and LOI results for starch-g-PNIPAAm copolymer and (starch-g-PNIPAAm)/EP BNCs

Sample code	TTI (s)	PHRR ($kW\ m^{-2}$)	THR ($MJ\ m^{-2}$)	MLR ($g\ s^{-1}$)	PSPR ($m^2\ s^{-1}$)	Char residue (wt %)	LOI (%)
S4	72±1	738±15	68±2	0.048±0.003	0.329±0.011	3.9±0.1	21.7
BNC1	69±2	613±11	61±2	0.041±0.005	0.171±0.003	6.2±0.1	24.2
BNC2	67±1	576±17	58±1	0.039±0.002	0.158±0.007	8.4±0.2	26.1
BNC3	65±1	534±11	54±1	0.035±0.001	0.127±0.023	11.9±0.1	28.4
BNC4	63±2	492±12	50±1	0.030±0.002	0.081±0.005	15.1±0.1	30.3

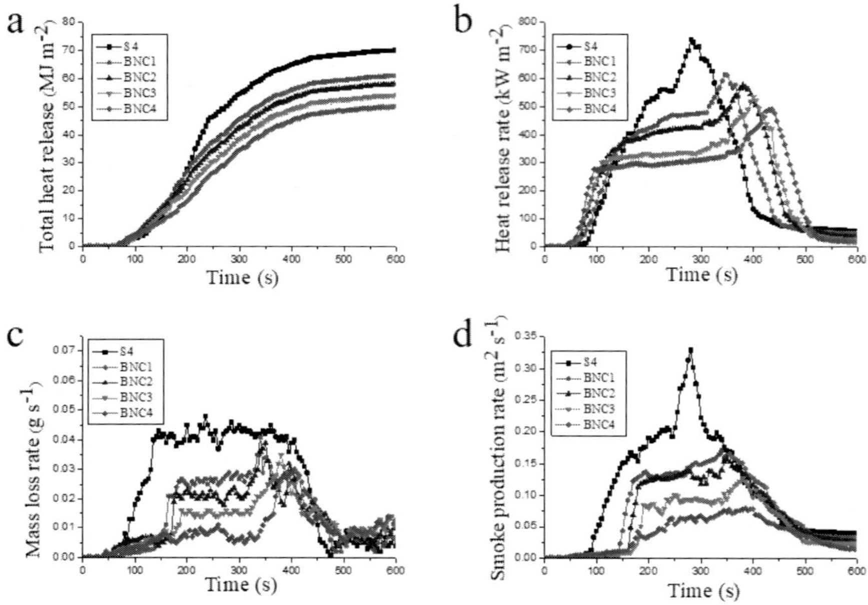

Figure 6. (a) THR, (b) HRR, (c) MLR and (d) SPR curves of starch-g-PNIPAAm copolymer and (starch-g-PNIPAAm)/EP BNCs.
[Copyright (2019), with permission from Elsevier [Details given in Reference No. 71]].

Apparently, neat S4 shows a TTI of 72±1 s, burns very fast after ignition, exhibiting a PHRR as high as 738±15 kW m-2 (Figure 6b), with generation of THR 68±2 MJ m-2 (Figure 6a). The EP addition reduced the TTI of BNCs is the result of catalytic effect of thermal decomposition for BNCs in comparison to S4 at initial stage, was in accord with earlier reports [69]. Interestingly, the TTI sharply reduces by a value of 9 s for BNC4 compared with that of S4. Meanwhile, both THR and PHRR of BNCs significantly reduce with EP incorporation, for instance, a THR and PHRR of 50±1 MJ m-2 and 492±12 kW m-2 respectively for BNC4 were observed. This reveals that 4% w/v EP incorporation enables the THR and PHRR to be reduced considerably by 26.5% and 33.3%, respectively, in relation to that of S4 (Figure 6a, b), strongly suggesting the remarkable enhancement in fire retardant property of the synthesised materials. This is the result of decarbonation of EP based $CaCO_3$ which releases non-combustible CO_2 gas on heating, involved in trapping degradation products into the carbonaceous char lead to reduction in heat transfer towards the materials (Figure 7) [65]. Similar phenomena were found in eggshell modified other polymeric systems [43].

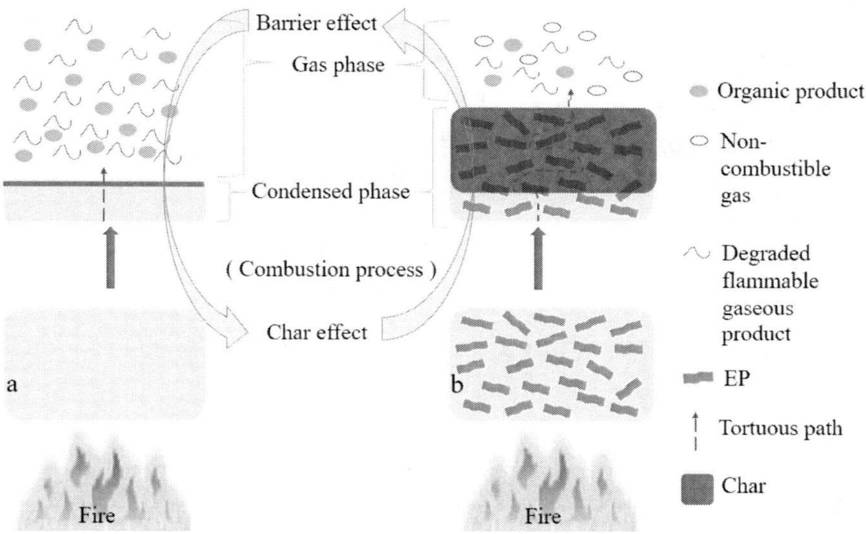

Figure 7. Mechanistic scheme of barrier and char effect for fire retardancy of (a) starch-g-PNIPAAm copolymer and (b) (starch-g-PNIPAAm)/EP BNC. [Copyright (2019), with permission from Elsevier [Details given in Reference No. 71]].

In addition, MLR of the materials also exhibits the similar trend as PHRR. MLR for S4 was found to be 0.048±0.003 g s-1, whereas MLR of BNCs was monotonously reduced to 0.041±0.005, 0.039±0.002, 0.035±0.001 and 0.030±0.002 g s-1 corresponding to 1%, 2%, 3%, 4% w/v of EP loadings respectively, as found in Figure 5c and Table 3. The significant decrease in MLR confirms the slowdown of combustion rate of BNCs in the presence of EP based CaCO3. Likewise, the concentration of char residual after cone calorimetry test was found to be increased monotonously with EP addition and reaches to 15.1±0.1 wt % for BNC4. Char residue with such high concentrations may act as a protective barrier and provides thermal insulation towards underlying materials, thereby resulting enhancement in fire retardancy of modified BNCs (Figure 7) [57].

Furthermore, this can be supported by the relation between the decrease in PHRR and increase in char residue. The peak SPR of the materials with 0%, 1%, 2%, 3% and 4% w/v EP contents were found to be 0.329±0.011, 0.171±0.003, 0.158±0.007, 0.127±0.023 and 0.081±0.005 m2 s-1 respectively as provided by Figure 5d and Table 3 revealing the effective dispersion of EP into the matrix by promoting char formation [69]. LOI test was performed for the further investigation of fire retardancy of BNCs. The LOI of BNCs increased with EP incorporation (Table 3). In comparison to S4 the LOI for BNC4 increased from 21.7% to 30.3% showing enhanced fire retardancy.

Analysis of Char Residue

After cone calorimeter tests the morphology as well as chemical compositions for char residuals are analysed, which can contribute for the understanding of flammability of EP modified BNCs in comparison to that of neat S4 matrix (Figure 8). A discontinuous structure with voids or holes and even large cracks are exhibited by the char residual obtained for S4 (Figure 8a). From the EDX results 58.16 wt% of carbon, 40.24 wt% of oxygen and 1.60 wt% of nitrogen are detected in char residuals of S4 (Figure 8a). BNC leaves a continuous and more compact char, which produces an insulating barrier as can be noticed in Figure 6b. This reveals an enhanced fire retardancy by acting as a thermal insulator and barrier to the degraded products (flammable gaseous products and organic products) at the time of combustion [69]. Moreover, besides 62.27 wt% of carbon, 30.73 wt% of oxygen, 1.99 wt% of nitrogen about 5.01 wt% of calcium was also found in the char residual of BNC4 (Figure 8b), that provides an improved resistance towards fire and is in accordance with the results by Xu et al. [57] and Yew et al. [65].

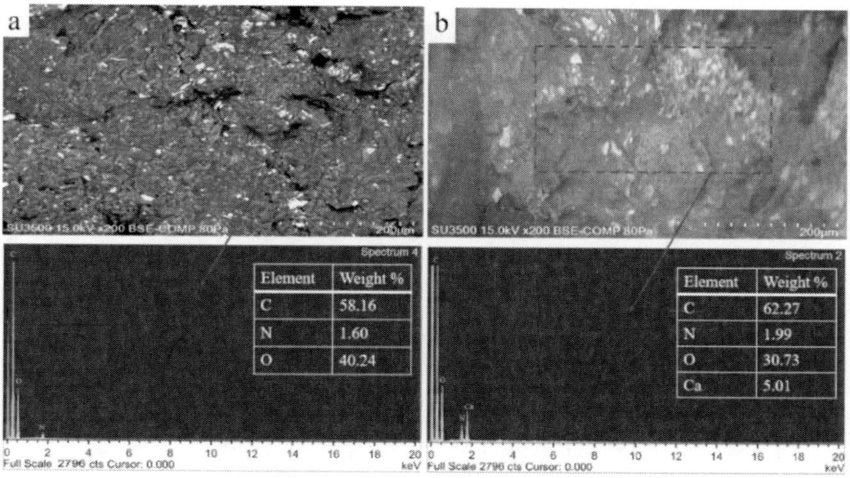

Figure 8. FESEM images and EDX mapping patterns of the char residues after cone calorimeter test: (a) S4 and (b) BNC4.
[Copyright (2019), with permission from Elsevier [Details given in Reference No. 71]].

Oxygen Barrier Property

The formulation of BNCs with low oxygen transmission rate is primary required for their industrialization. The EP incorporation to polymer matrix is an alternate method to reduce its oxygen transmission [56]. The influence of EP on oxygen transmission of prepared BNCs was shown in Figure 9. It was noticed that OTR through all the BNCs was less compared to S4 matrix. Furthermore OTR for BNC4 found to decrease with EP loading and significantly reduced by 73% in comparison to S4 (Figure 9). This might be the result of homogeneous dispersion of EP throughout starch-g-PNIPAAm matrix and densification of polymer network. Also, EP incorporation into the matrix would generate a tortuous path for OP, leading to enhanced oxygen barrier property (Figure 7) [32, 56].

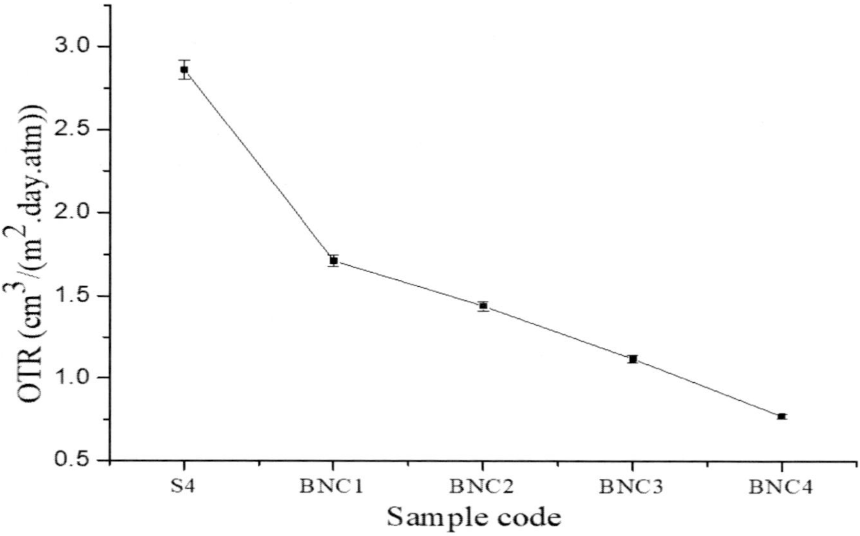

Figure 9. Oxygen transmission rate of BNCs as a function of EP contents. [Copyright (2019), with permission from Elsevier [Details given in Reference No. 71]].

Biodegradation

The biodegradation of EP incorporated BNCs was compared with virgin starch-g-PNIPAAm matrix for 180 days with regular interval of 30 days. This technique includes the calculation of percentage weight loss of prepared materials in sludge water. The decrease in percentage weight loss of BNCs with increase in EP content (or decrease in starch content) was noticed as provided in Figure 10. However, biodegradation of BNCs and virgin copolymer was improved with time duration. The uniform dispersion of hydrophobic bio-filler EP with virgin starch-g-PNIPAAm matrix not only increase the hydrophobicity of BNCs but also hindered the absorption of water and microorganisms by BNCs result in decrease of degradation rate [30]. Again enzymes released by microorganisms, reason for the biodegradation of starch by hydrolysis, along with their activity were reduced with decrease in starch content [45].

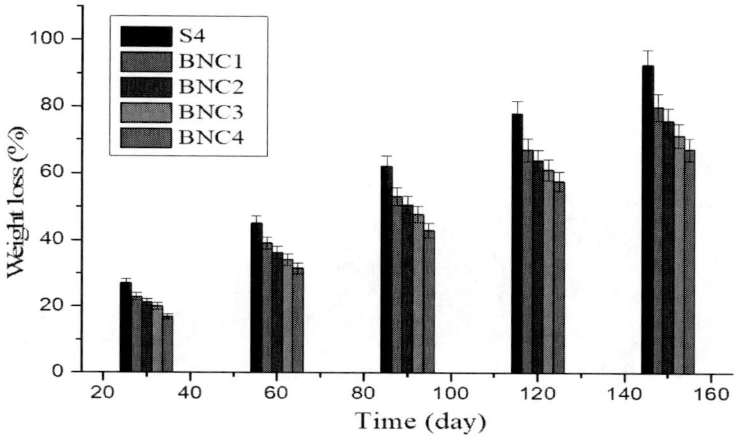

Figure 10. Biodegradation of starch-g-PNIPAAm copolymer and (starch-g-PNIPAAm)/EP BNCs.
[Copyright (2019), with permission from Elsevier [Details given in Reference No. 71]].

CONCLUSION

This work is meant for designing new-generation polysaccharide based fire retardant NCs to reduce excessive consumption of the petroleum products and opened a huge window of opportunities for polymeric NCs. A new-generation (starch-g-PNIPAAm)/EP BNCs have been developed via emulsifier-free emulsion technique. The miscibility and uniform dispersion of EP based CaCO3 within starch-g-PNIPAAm matrix were resulted an increase in YM of 4.6±0.1 GPa and TS of 67±1 MPa for BNC (4% w/v EP). Compared to neat starch-g-PNIPAAm, inclusion of EP (4% w/v) exhibited remarkable suppression of fire hazards of (starch-g-PNIPAAm)/EP BNC in terms of reduced THR (decreased by 26.5%) and PHRR (decreased by 33.3%) as well as formation of 15.1±0.1 wt % of char residue. TGA experiment showed the improved thermal stability for the synthesized BNCs in comparison to starch-g-PNIPAAm matrix. In addition, BNC with a clay concentration of 4% w/v exhibits an OTR of 0.77 cm3/(m2.day.atm). More importantly, this work highlights a sustainable and green methodology to design non-halogenated fire hazards suppression starch based biomaterials to reach the constantly changing demands of new regulations, test methods and standards for wide industrial applications via a cost-competitive method for wide industrial applications.

REFERENCES

[1] Dittenber, D. B., & Ganga Rao, H. V. (2012). Critical review of recent publications on use of natural composites in infrastructure. *Composites Part A: applied science and manufacturing*, 43(8), 1419-1429. Oprisan, G., N. Taranu, and V. Munteanu. "Buletinul Institutului Politehnic Diniasi." LVI (LX) 3 (2010).

[2] Morgan, A. B., & Wilkie, C. A. (2007). *Practical issues and future trends in polymer nanocomposite flammability research*, 362-372. John Wiley & Sons, New York, NY, USA.

[3] Camino, G., Costa, L., & Di Cortemiglia, M. L. (1991). Overview of fire retardant mechanisms. *Polymer Degradation and Stability*, *33*(2), 131-154.

[4] Wilkie, C. A., & Morgan, A. B. (2009). *Fire retardancy of polymeric materials*. CRC press.

[5] Woodrow, B. (2012). Fire as vulnerability: the value added from adopting a vulnerability approach. *World Fire Statistics Bulletin*, Valéria Pacella.

[6] Karter, M. J., & Molis, J. L. (2014). *US firefighter injuries-2013*, 1138-1145.

[7] Ha-Duong, M., & Journé, V. (2014). Calculating nuclear accident probabilities from empirical frequencies. *Environment Systems and Decisions*, 34(2), 249-258.

[8] Marker, T. R., Sarkos, C. P., & Hill, R. G. (1996, October). Water spray system development and evaluation for enhanced postcrash fire survivability and in-flight protection in cargo compartments. *In 88th Symp. of the Propulsion and Energetics Panel on Aircraft Fire Safety*.

[9] Mangiacapra, P., Gorrasi, G., Sorrentino, A., & Vittoria, V. (2006). Biodegradable nanocomposites obtained by ball milling of pectin and montmorillonites. *Carbohydrate polymers*, 64(4), 516-523.

[10] Gontard, N., & Guilbert, S. (1994). Bio-packaging: technology and properties of edible and/or biodegradable material of agricultural origin. In *Food packaging and preservation*, 159-181. Springer, Boston, MA.

[11] Cheng, S., Zhang, Y., Cha, R., Yang, J., & Jiang, X. (2016). Water-soluble nanocrystalline cellulose films with highly transparent and oxygen barrier properties. *Nanoscale*, 8(2), 973-978.

[12] Ali, S. S., Tang, X., Alavi, S., & Faubion, J. (2011). Structure and physical properties of starch/poly vinyl alcohol/sodium montmorillonite nanocomposite films. *Journal of agricultural and food chemistry*, 59(23), 12384-12395.

[13] Makhijani, K., Kumar, R., & Sharma, S. K. (2015). Biodegradability of blended polymers: A comparison of various properties. *Critical Reviews in Environmental Science and Technology*, 45(16), 1801-1825.

[14] Rasal, R. M., Janorkar, A. V., & Hirt, D. E. (2010). Poly (lactic acid) modifications. *Progress in polymer science*, 35(3), 338-356.

[15] Shumao, L., Jie, R., Hua, Y., Tao, Y., & Weizhong, Y. (2010). Influence of ammonium polyphosphate on the flame retardancy and mechanical properties of ramie fiber-reinforced poly (lactic acid) biocomposites. *Polymer International*, 59(2), 242-248.

[16] Zou, G. X., Jin, P. Q., & Xin, L. Z. (2008). Extruded starch/PVA composites: water resistance, thermal properties, and morphology. *Journal of Elastomers & Plastics*, 40(4), 303-316.

[17] Tang, X., & Alavi, S. (2011). Recent advances in starch, polyvinyl alcohol based polymer blends, nanocomposites and their biodegradability. *Carbohydrate polymers*, 85(1), 7-16.

[18] Siqueira, G., Bras, J., & Dufresne, A. (2009). Cellulose whiskers versus microfibrils: influence of the nature of the nanoparticle and its surface functionalization on the thermal and mechanical properties of nanocomposites. *Biomacromolecules*, 10(2), 425-432.

[19] Swain, S. K., Patra, S. K., & Kisku, S. K. (2014). Study of thermal, oxygen-barrier, fire-retardant and biodegradable properties of starch bionanocomposites. *Polymer composites*, 35(7), 1238-1243.

[20] Belgacem, M. N., & Gandini, A. (2005). The surface modification of cellulose fibres for use as reinforcing elements in composite materials. *Composite Interfaces*, 12(1-2), 41-75.

[21] Orts, W. J., Shey, J., Imam, S. H., Glenn, G. M., Guttman, M. E., & Revol, J. F. (2005). Application of cellulose microfibrils in polymer nanocomposites. *Journal of Polymers and the Environment*, 13(4), 301-306.

[22] López, J. P., Mutjé, P., Carvalho, A. J. F., Curvelo, A. A. S., & Gironès, J. (2013). Newspaper fiber-reinforced thermoplastic starch biocomposites obtained by melt processing: Evaluation of the mechanical, thermal and water sorption properties. *Industrial Crops and Products*, 44, 300-305.

[23] Ke, Y., Long, C., & Qi, Z. (1999). Crystallization, properties, and crystal and nanoscale morphology of PET–clay nanocomposites. *Journal of Applied Polymer Science*, 71(7), 1139-1146.

[24] Zeng, Q. H., Yu, A. B., Lu, G. Q., & Paul, D. R. (2005). Clay-based polymer nanocomposites: research and commercial development. *Journal of nanoscience and nanotechnology*, 5(10), 1574-1592.

[25] Samal, R., & Sahoo, P. K. (2009). Development of a biodegradable rice straw-g-poly (methyl methacrylate)/sodium silicate composite flame retardant. *Journal of applied polymer science*, 113(6), 3710-3715.

[26] Sahu, M., & Sahoo, P. K. (2015). *Synthesis and characterization of flame retardant jute-g-poly (2-ethyl hexylacrylate)/montmorillonite composites.*

[27] Mallick, B. N., Rana, P. K., & Sahoo, P. K. (2015). Preparation of Starch-Based Bionanocomposite Hydrogel with Mica for Fire Retardancy. *Advances in Polymer Technology*, 34(4).

[28] Jena, D. K., & Sahoo, P. K. (2018). Development of biodegradable cellulose-g-poly (butyl acrylate)/kaolin nanocomposite with

improved fire retardancy and mechanical properties. *Journal of Applied Polymer Science*, 135(11), 45968.
[29] Bootklad, M., & Kaewtatip, K. (2013). Biodegradation of thermoplastic starch/eggshell powder composites. *Carbohydrate polymers*, 97(2), 315-320.
[30] Zhang, Q. X., Yu, Z. Z., Xie, X. L., Naito, K., & Kagawa, Y. (2007). Preparation and crystalline morphology of biodegradable starch/clay nanocomposites. *Polymer*, 48(24), 7193-7200.
[31] Cheng, S., Zhang, Y., Cha, R., Yang, J., & Jiang, X. (2016). Water-soluble nanocrystalline cellulose films with highly transparent and oxygen barrier properties. *Nanoscale*, 8(2), 973-978.
[32] Carosio, F., Fontaine, G., Alongi, J., & Bourbigot, S. (2015). Starch-based layer by layer assembly: efficient and sustainable approach to cotton fire protection. *ACS applied materials & interfaces*, 7(22), 12158-12167.
[33] Ray, S. S., & Bousmina, M. (2005). Biodegradable polymers and their layered silicate nanocomposites: in greening the 21st century materials world. *Progress in materials science*, 50(8), 962-1079.
[34] Dean, K. M., Do, M. D., Petinakis, E., & Yu, L. (2008). Key interactions in biodegradable thermoplastic starch/poly (vinyl alcohol)/montmorillonite micro-and nanocomposites. *Composites Science and Technology*, 68(6), 1453-1462.
[35] Wang, N., Zhang, X., Han, N., & Bai, S. (2009). Effect of citric acid and processing on the performance of thermoplastic starch/montmorillonite nanocomposites. *Carbohydrate polymers*, 76(1), 68-73.
[36] Chen, H., Shan, Z. H., Woo, M. W., & Chen, X. D. (2017). Preparation and characteristic of gelatine/oxidized corn starch and gelatin/corn starch blend microspheres. *International journal of biological macromolecules*, 94, 326-334.
[37] Da Róz, A. L., Zambon, M. D., Curvelo, A. A., & Carvalho, A. J. (2011). Thermoplastic starch modified during melt processing with organic acids: The effect of molar mass on thermal and mechanical properties. *Industrial Crops and Products*, 33(1), 152-157.

[38] Van Krevelen, D. W., & Te Nijenhuis, K. (2009). *Properties of polymers: their correlation with chemical structure; their numerical estimation and prediction from additive group contributions*. Elsevier.

[39] Dean, K., Yu, L., & Wu, D. Y. (2007). Preparation and characterization of melt-extruded thermoplastic starch/clay nanocomposites. *Composites Science and Technology*, 67(3-4), 413-421.

[40] Tang, X., Alavi, S., & Herald, T. J. (2008). Barrier and mechanical properties of starch-clay nanocomposite films. *Cereal chemistry*, 85(3), 433-439.

[41] Tang, X., Alavi, S., & Herald, T. J. (2008). Effects of plasticizers on the structure and properties of starch–clay nanocomposite films. *Carbohydrate Polymers*, 74(3), 552-558.

[42] Nascimento, T. A., Calado, V., & Carvalho, C. W. P. (2012). Development and characterization of flexible film based on starch and passion fruit mesocarp flour with nanoparticles. *Food Research International*, 49(1), 588-595.

[43] Wang, X., Hu, Y., Song, L., Xuan, S., Xing, W., Bai, Z., & Lu, H. (2011). Flame retardancy and thermal degradation of intumescent flame retardant poly (lactic acid)/starch biocomposites. *Industrial & engineering chemistry research*, 50(2), 713-720.

[44] Hanifi, S., Oromiehie, A., Ahmadi, S., & Farhadnejad, H. (2014). (Corn Starch and Montmorillonite Nanocomposite)-Reinforced Polypropylene: Preparation, Properties, and Biodegradability. *Journal of Vinyl and Additive Technology*, 20(1), 16-23.

[45] Dupretz, R., Fontaine, G., & Bourbigot, S. (2014). Fire retardancy of a new polypropylene-grafted starch: Part II: investigation of mechanisms. *Journal of Fire Sciences*, 32(3), 210-229.

[46] Majdzadeh-Ardakani, K., Navarchian, A. H., & Sadeghi, F. (2010). Optimization of mechanical properties of thermoplastic starch/clay nanocomposites. *Carbohydrate Polymers*, 79(3), 547-554.

[47] Haraguchi, K., Takehisa, T., & Fan, S. (2002). Effects of clay content on the properties of nanocomposite hydrogels composed of poly (N-

isopropylacrylamide) and clay. *Macromolecules*, 35(27), 10162-10171.

[48] Tian, B. S., & Yang, C. (2009). Temperature-responsive nanocomposites based on mesoporous SBA-15 silica and PNIPAAm: synthesis and characterization. *The Journal of Physical Chemistry C*, 113(12), 4925-4931.

[49] Ma, J., Zhang, L., Fan, B., Xu, Y., & Liang, B. (2008). A novel sodium carboxymethylcellulose/poly (N-isopropylacrylamide)/Clay semi-IPN nanocomposite hydrogel with improved response rate and mechanical properties. *Journal of Polymer Science Part B: Polymer Physics*, 46(15), 1546-1555.

[50] Gil, E. S., Park, S. H., Tien, L. W., Trimmer, B., Hudson, S. M., & Kaplan, D. L. (2010). Mechanically robust, rapidly actuating, and biologically functionalized macroporous poly (N-isopropylacrylamide)/silk hybrid hydrogels. *Langmuir*, 26(19), 15614-15624.

[51] Wang, Q., Asoh, T. A., & Uyama, H. (2018). Rapid uniaxial actuation of layered bacterial cellulose/poly (N-isopropylacrylamide) composite hydrogel with high mechanical strength. *RSC advances*, 8(23), 12608-12613.

[52] Chang, C., Han, K., & Zhang, L. (2011). Structure and properties of cellulose/poly (N-isopropylacrylamide) hydrogels prepared by IPN strategy. *Polymers for Advanced Technologies*, 22(9), 1329-1334.

[53] Tan, Y., Xu, K., Wang, P., Li, W., Sun, S., & Dong, L. (2010). High mechanical strength and rapid response rate of poly (N-isopropyl acrylamide) hydrogel crosslinked by starch-based nanospheres. *Soft Matter*, 6(7), 1467-1471.

[54] Rahman, G. S., Aftab, H., Islam, M. S., Mukhlish, M. Z. B., & Ali, F. (2016). Enhanced physico-mechanical properties of polyester resin film using CaCO 3 filler. *Fibers and Polymers*, 17(1), 59-65.

[55] Jiang, B., Li, S., Wu, Y., Song, J., Chen, S., Li, X., & Sun, H. (2018). Preparation and characterization of natural corn starch-based composite films reinforced by eggshell powder. *CyTA-Journal of Food*, 16(1), 1045-1054.

[56] Xu, Z., Chu, Z., Yan, L., Chen, H., Jia, H., & Tang, W. (2019). Effect of chicken eggshell on the flame-retardant and smoke suppression properties of an epoxy-based traditional APP-PER-MEL system. *Polymer Composites*, 40(7), 2712-2723.

[57] Pradhan, A. K., & Sahoo, P. K. (2017). Synthesis and study of thermal, mechanical and biodegradation properties of chitosan-g-PMMA with chicken egg shell (nano-CaO) as a novel bio-filler. *Materials Science and Engineering: C*, 80, 149-155.

[58] Kang, D. J., Pal, K., Park, S. J., Bang, D. S., & Kim, J. K. (2010). Effect of eggshell and silk fibroin on styrene–ethylene/butylene–styrene as bio-filler. *Materials & Design*, 31(4), 2216-2219.

[59] Lin, Z., Zhang, Z., & Mai, K. (2012). Preparation and properties of eggshell/β-polypropylene bio-composites. *Journal of applied polymer science*, 125(1), 61-66.

[60] Prabhakar, M. N., Rehaman Shah, A. U., & Song, J. I. (2016). Fabrication and characterization of eggshell powder particles fused wheat protein isolate green composite for packaging applications. *Polymer Composites*, 37(11), 3280-3287.

[61] Xu, Y., & Hanna, M. (2007). Effect of eggshell powder as nucleating agent on the structure, morphology and functional properties of normal corn starch foams. *Packaging Technology and Science: An International Journal*, 20(3), 165-172.

[62] Yoo, S., Hsieh, J. S., Zou, P., & Kokoszka, J. (2009). Utilization of calcium carbonate particles from eggshell waste as coating pigments for ink-jet printing paper. *Bioresource technology*, 100(24), 6416-6421.

[63] Habeeb, O. A., Yasin, F. M., & Danhassan, U. A. (2014). Characterization and application of chicken eggshell as green adsorbents for removal of H2S from wastewaters. *IOSR J Environ Sci Toxicol Food Technol*, 8(11), 7-12.

[64] Yew, M. C., Sulong, N. R., Yew, M. K., Amalina, M. A., & Johan, M. R. (2015). Eggshells: A novel bio-filler for intumescent flame-retardant coatings. *Progress in Organic Coatings*, 81, 116-124.

[65] Sugawara-Narutaki, A. (2013). Bio-inspired synthesis of polymer–inorganic nanocomposite materials in mild aqueous systems. *Polymer journal*, 45(3), 269-276.

[66] Mohan, T. P., & Kanny, K. (2018). Thermal, mechanical and physical properties of nanoegg shell particle-filled epoxy nanocomposites. *Journal of Composite Materials*, 52(29), 3989-4000.

[67] Hassen, A. A., Dizbay-Onat, M., Bansal, D., Bayush, T., & Vaidya, U. (2015). Utilization of chicken eggshell waste as a bio-filler for thermoplastic polymers: thermal and mechanical characterization of polypropylene filled with naturally derived CaCO3. *Polymers and Polymer Composites*, 23(9), 653-662.

[68] Jena, D. K., & Sahoo, P. K. (2018). Simultaneous improvement of mechanical and fire retardant properties of synthesised biodegradable guar gum-g-poly (butyl acrylate)/montmorillonite nanocomposite. *Polymer Degradation and Stability*, 154, 37-45.

[69] Sahoo, P. K., & Jena, D. K. (2018). Synthesis and study of mechanical and fire retardant properties of (carboxymethyl cellulose-g-polyacrylonitrile)/montmorillonite biodegradable nanocomposite. *Journal of Polymer Research*, 25(12), 260.

[70] Jena, D. K., & Sahoo, P. K. (2019). New strategies for the construction of eggshell powder reinforced starch based fire hazard suppression biomaterials with tailorable thermal, mechanical and oxygen barrier properties. *International journal of biological macromolecules*, 140, 496-504.

BIOGRAPHICAL SKETCHES

Prafulla Kumar Sahoo, Professor

Affiliation: Department of Chemistry, Utkal University, Vani Vihar, Bhubaneswar-751004, Odisha, India

Education: M.Sc., M.Phil., Ph.D., PDF (USA), C.Chem., MRSC

Business Address: Department of Chemistry, Utkal University, Vani Vihar, Bhubaneswar-751004, Odisha, India
Email: psahoochemuu@gmail.com

Research and Professional Experience: 35 years

Professional Appointments: Professor of Chemistry

Honors:

1. Distinguished Visiting Professor, Beijing Institute of Technology, Beijing, China.
2. Executive Council Member, Asia-Oceania Fire Safety Materials & Engineering (AOFSM), China.
3. Adviser, Regional Advisory Committee, CIPET, Odisha (Govt. of India).

Publications from the Last 3 Years:

1) Sethy, T. R., Pradhan, A. K., & Sahoo, P. K. (2019). Simultaneous studies on kinetics, bio-adsorption behaviour of chitosan grafted thin film nanohydrogel for removal of hazardous metal ion from water. *Environmental Nanotechnology, Monitoring & Management*, *12*, 100262.
2) Jena, D. K., & Sahoo, P. K. (2019). New strategies for the construction of eggshell powder reinforced starch based fire hazard suppression biomaterials with tailorable thermal, mechanical and oxygen barrier properties. *International Journal of Biological Macromolecules*, *140*, 496-504.
3) Sethy, T. R., & Sahoo, P. K. (2019). Highly toxic Cr (VI) adsorption by (chitosan-g-PMMA)/silica bionanocomposite prepared via emulsifier-free emulsion polymerisation. *International journal of biological macromolecules*, *122*, 1184-1190.

4) Sahoo, P. K., & Jena, D. K. (2018). Synthesis and study of mechanical and fire retardant properties of (carboxymethyl cellulose-g-polyacrylonitrile)/montmorillonite biodegradable nanocomposite. *Journal of Polymer Research*, *25*(12), 260-269.
5) Jena, D. K., & Sahoo, P. K. (2018). Simultaneous improvement of mechanical and fire retardant properties of synthesised biodegradable guar gum-g-poly (butyl acrylate)/montmorillonite nanocomposite. *Polymer Degradation and Stability*, *154*, 37-45.
6) Jena, D. K., & Sahoo, P. K. (2018). Development of biodegradable cellulose-g-poly (butyl acrylate)/kaolin nanocomposite with improved fire retardancy and mechanical properties. *Journal of Applied Polymer Science*, *135*(11), 45968-45975.
7) Pradhan, A. K., Sahoo, P. K., & Rana, P. K. (2017). Chitosan-g-PMMA/Kaolin Bionanocomposites for Use in Bioadhesive Bone-Cement Implants. *Journal of Renewable Materials*, *5*(5), 371-379.
8) Pradhan, A. K., & Sahoo, P. K. (2017). Synthesis and study of thermal, mechanical and biodegradation properties of chitosan-g-PMMA with chicken egg shell (nano-CaO) as a novel bio-filler. *Materials Science and Engineering: C*, *80*, 149-155.

Dharmendra Kumar Jena

Affiliation: Department of Chemistry, Utkal University, Vani Vihar, Bhubaneswar-751004, Odisha, India

Education: M.Sc., M.Phil., Ph.D. (Continuing)

Business Address: Department of Chemistry, Utkal University, Vani Vihar, Bhubaneswar-751004, Odisha, India
Email: dharmendrajena88@gmail.com

Research and Professional Experience: 6 years

Professional Appointments: Research Scholar (UGC-BSR -SRF, Govt. of India)

Publications from the Last 3 Years:

1) Jena, D. K., & Sahoo, P. K. (2019). New strategies for the construction of eggshell powder reinforced starch based fire hazard suppression biomaterials with tailorable thermal, mechanical and oxygen barrier properties. *International Journal of Biological Macromolecules*, 140, 496-504.
2) Sahoo, P. K., & Jena, D. K. (2018). Synthesis and study of mechanical and fire retardant properties of (carboxymethyl cellulose-g-polyacrylonitrile)/montmorillonite biodegradable nanocomposite. *Journal of Polymer Research*, 25(12), 260-269.
3) Jena, D. K., & Sahoo, P. K. (2018). Simultaneous improvement of mechanical and fire retardant properties of synthesised biodegradable guar gum-g-poly (butyl acrylate)/montmorillonite nanocomposite. *Polymer Degradation and Stability*, 154, 37-45.
4) Jena, D. K., & Sahoo, P. K. (2018). Development of biodegradable cellulose-g-poly (butyl acrylate)/kaolin nanocomposite with improved fire retardancy and mechanical properties. *Journal of Applied Polymer Science*, 135(11), 45968-45975.

In: Critical Issues and Analysis … ISBN: 978-1-53618-738-0
Editors: A. Cablé and B. Manescau

Chapter 5

FIRE PROTECTION OF CULTURAL HERITAGE: EVALUATION OF AN INNOVATIVE VACUUM SPRINKLER SYSTEM

***Khaled Chetehouna*[1], *Axel Cablé*[1,2*],**
***Shehu A. Abdulrahman*[1] *and Maurice Kadoche*[2]**
[1]INSA Centre Val de Loire, PRISME EA 4229, Bourges, France
[2]EXEC, Goussainville, France

ABSTRACT

The present study concerns the efficient design of full-scale automatic sprinkler systems for the fire protection of historic buildings (castles, cathedrals, museums…). In fact, fire suppression is rendered difficult in heritage buildings because of their specific architectural features, high flammability of materials, and vulnerability to water leaks and false activations. An inventory of the requirements and current state of protection of heritage buildings is first presented in this chapter, highlighting the potential shortcomings of conventional fire protection systems. Then, an innovative fire sprinkler system whose piping is kept

* Corresponding Author's Email: axel.cable@ibs-conseil.fr.

under negative relative pressure at idle state (vacuum) is introduced, and evaluated as compared to conventional sprinkler systems filled with pressurized air (dry pipe). To this end, specific routines (sprinkler, fire pump, etc.) have been introduced within the OpenFOAM Computational Fluid Mechanics (CFD) package. Simulations of the two-phase compressible transient air/water flow within the pipes subsequently to the opening of a sprinkler head were then carried out, and validated by means of measurements on a scale 1 test bench. This approach allowed to compare the fire performance of both sprinkler systems systems in terms of: *(i)* activation time, *(ii)* delay necessary for water delivery after activation and *(iii)* water distribution in the pipes, for different scenarios related to the fire safety of historic buildings.

Keywords: fire safety, sprinkler systems, cultural heritage, full-scale experiments, CFD simulations

INTRODUCTION

Historic buildings such as castles, cathedrals, museums, archives… are exposed to a significant risk of fire, with potentially severe consequences and destruction of precious and irreplaceable heritage (artefacts, works of art, furniture...). In fact, fire suppression is rendered difficult in heritage buildings because of their specific architectural features (high ceilings, long corridors increasing fire spread, beamed ceilings and molding hindering the sprinkler sprays, etc.) and high flammability of materials (tapestries, woodwork…) and constraints (esthetics, restricted access to water sources…). Furthermore, compliance to fire standards is rarely achieved.

As a consequence, hundreds of historic buildings have been completely destroyed by fire and serious losses still occur every year. The recent Notre-Dame Cathedral fire [1] in Paris in 2019 (Figure 1) is a reminder of how centuries of history can be lost in few minutes. Apart from damaging the building, fire often destroys irreplaceable collections. Only through effective fire protection methods will our built history will be preserved for future generations. These buildings require automatic fire suppression systems because, if a fire starts, very often the intervention of the fire-fighters occurs too late. This is what happened in September 2018 in the 200-year-old

National Museum in Rio de Janeiro – Brazil (Figure 2). The fire began after it closed to the public on Sunday and raged into the night. There were no reports of injuries, but the loss to Brazilian science, history and culture was incalculable [2]. Other notorious examples include La Fenice theater in Venice (Italy) in 1996 and the Windsor Castle (England) in 1992.

Figure 1. Notre-Dame Cathedral fire, Paris. April 15-16, 2019 [5].

The causes of the fires within these buildings are multiple: the main cause is of electrical nature and results, on the one hand, from maintenance works carried out by external contractors on the buildings, and on the second hand, from obsolete electrical installations in combination with rodent damage resulting in short-circuiting [3]. The second main cause of fire is localized in the attics of these historic buildings, which are rarely visited, and results from the accumulation of dust and fermentation of bird and rodent excrements. Self-combustion then occurs, and results in the beginning of a fire.

Figure 2. Fire at the National Museum of Brazil, Rio de Janeiro. September 2-3, 2018 [6].

Regarding the characteristics of the fire, onsite inspections and research have shown that in many cases, fires for such buildings start of as compartment fires, but then quickly extends to the frame and to the roof with, as a consequence, a large overflow of flames and plume developing outside of the building. This contributes to the massive and destructive nature of these disasters, which present characteristics similar to those of massive tall-buildings fires or wildland fires [4].

Use of Automatic Fire Protection Systems in Cultural heritage

Nowadays, automatic fire protection systems is considered as one of the best strategy to decrease the consequences of fires in buildings. They are mainly two types of principles available: water spray (sprinkler) and water mist. The former appeal to large drops (with a diameter of the order of a millimeter or more) and suppresses or contains the fire by cooling the fire's adjacent surfaces and air temperature, and the latter with a finer particle size (typically less than 400 μm), also acting on the displacement of oxygen.

The American NFPA has estimated that the use of sprinklers in heritage buildings enables to reduce the damage as much as 79% in libraries, 92% in museums and 67% in religious facilities, if the systems are completely operative. Nevertheless, the fight against fire hazard in buildings is complicated by failures in fire sprinkler systems (up to 9%) [7].

In general, the main elements of any fire protection system are a water source, a control system with pumps, one or several piping sections that cover the different areas to be protected and the sprinklers that spray water in case of fire. The state-of-the-art fire suppression systems can be classified in two basic types:

- Wet systems: the piping network is permanently filled with water (+9.5 bar relative pressure).
- Dry systems: the piping network contains pressurised air (+2.5 bar relative pressure).

The experience demonstrates that, in both cases, malfunction or failures are due to obstruction in the sprinkler piping network, a problem usually generated when the systems are in supervisory mode (over 99% of the time). In fact, over time, sludge and corrosion can cause water leaks that can damage property and the building, as well as false, delayed or failed activation of the fire suppression systems.

Each time a wet pipe system is drained and refilled, the corrosion rate accelerates as dioxygen is introduced into the network (the higher the temperature and pressure, the faster the corrosion rate). This problem is accelerated if non drinkable water is used or the installation is in aggressive climate areas. Moreover, water and moisture offer biotopes favouring the development of different forms of life (bacteria, microorganisms, etc.), generating microbiologically induced corrosion (MIC) and slush. Dry pipe systems are even more prone to corrosion than wet systems because they contain pressurised air and therefore a higher quantity of O_2. The compressed air itself introduces moisture into the system and, periodically, these systems are hydrostatically pressure-tested to make sure they do not leak, but taking all the moisture out is impossible.

As a result, over 35% of wet systems and 73% of dry systems have serious corrosion issues (pipes that require replacement) [7]. Corrosion leads to catastrophic failures in the piping network and the sprinklers of fire protection systems.

- The formation of pinholes results in water leaks in wet systems, leading to water flow reduction and generating important property damages.
- Corrosion reduces the pipe section, hindering the water flow (20% reduction means 40% less water flow) and generating higher roughness and pressure losses inside the pipes. Rust debris can even completely block the water supply and obstruct the sprinkler heads, making the system fail.

Non-visible failures can completely disable a fire suppression system without being detected. Although the state-of-the-art systems have to be subjected to constant inspections and testing to ensure proper functioning, consuming time and money, these situations are very common [8].

Another problem is the risk of freezing resulting in ice plugs blocking the flow and in broken pipes leading to water damage. Even the dry systems can suffer from this problem because the moisture of the compressed air can condense and freeze in the pipes too. This risk is evident in countries where freezing temperatures are common in winter, but also in unheated areas such as attics, which can be frequent in historic monuments such as castles or cathedrals. In fact, buildings located in places where temperatures may fall below 4°C [9] have to be permanently heated just to avoid freezing of the fire sprinkler system.

Finally, sprinkler systems require the installation of pipes in the walls of the buildings, and sprinkler heads or nozzles at ceiling, which are considered anaesthetic and most often require approval of specific preservation architects which must preserve the criteria aesthetics of the building, depending on the countries. The systems also need to be frequently maintained and drained. All of these parameters mean that building owners often prefer not to use fire protection, and as a consequence, automatic fire extinguishing systems based on sprinklers or mists of water, are to date rarely installed in the context of the protection of these monuments. Indeed, the risks of water leaks and water damage which could potentially damage the irreplaceable cultural heritage such as works of art, tapestries, furniture are significant...

Thus, the prevention of fires in the cultural and built heritage (churches, cathedrals, castles, museums, archives, etc.) is generally limited to the implementation of a fire detection devices (smoke detector or infra-red temperature sensors) aiming to alert the local fire station and aiming for a relatively quick intervention of firefighters following the departure of the fire. Unfortunately, given the large amount of combustible materials in historic buildings, it is often too late to prevent severe damage.

To give an example of the current state of protection in Europe, the Confederation of Fire Protection Associations Europe (CFPA-Europe) is an association of national organisations concerned primarily with fire prevention and protection and also with safety, security and other associated risks. CFPA-Europe has issued guidelines regarding fire protection in different applications. Guideline No 30:2013 F is specifically devoted to the protection of historic buildings [10]. In Norway in particular, the Directorate for Cultural Heritage "Riksantikvaren" is the responsible for the management of all archaeological and architectural monuments and sites and cultural environment. Riksantikvaren, directly responsible for the fire protection of 28 old wooden churches (some of them built more than 800 years ago), recommends that all Norwegian churches made of wood and those made of stone or bricks with wooden roofs, should have automatic fire extinguishing systems (sprinkler or water mist).

Requirements

Hence, in order to increase the efficiency, reliability and sustainability of fire protection systems, it is necessary to find new technical solutions capable to overcome the problems and weaknesses of the conventional automatic fire protection, some of which are summarized below:

- High risk of malfunction or failure due to corrosion deposits;
- Risk of water damage in case of pin holes or accidental activation;
- Contamination caused by products added to the water to prevent freezing or corrosion;
- Only drinkable water can be used;
- Risk of freezing in cold environments;
- Longer water delivery time and insufficient flow/pressure;
- Need of regular control and maintenance.

Figure 3. Illustration of a fire sprinkler system in a castle of the Loire Valley, France, with proper exterior (facade) and interior (ceiling) integration.

Heritage buildings require tailored solutions for their fire protection systems. These solutions have to be minimally visible as these buildings can have high ornate ceilings with moulded cornices or wooden wall panelling

(Figure 3). The challenge is to provide an appropriate fire protection system with minimum impact on the building.

In this context and regarding the available scientific literature, relatively few studies are devoted to the fire safety of cultural heritage. Most available studies concern the development of fire risk assessment methods and response plan for heritage buildings [11-15], and engineering principles for the fire safety of historic buildings [16, 17]. Other specific studies are devoted to passive strategies for the fire safety of museums and libraries [2], the assessment of the fire performance of contemporary timber and new materials [18, 19], the development of flame retardants for wooden heritage buildings [20] or the impact of fire extinguisher agents on cultural resource materials [21].

However, no studies were found regarding the evaluation of the performance of innovative fire suppression systems for the protection of cultural heritage, except for a study on hypoxic air venting for the preservation of heritage documents and works of art [22].

Potential solution: Vacuum Sprinkler System

In this context, the present chapter aims to study the performance of an innovative automatic fire protection system whose pipes are placed and maintained at a negative relative pressure relatively to the atmospheric pressure (p = -0.18 bar rel.) at idle state, i.e., a *vacuum sprinkler system*. By removing water and most of the oxygen in the pipes at idle state, this system aims to solve all problems relative to conventional sprinkler systems (leaks, risk of freezing, corrosion, etc.). The way the vacuum sprinkler system operates from supervisory state to the suppression of a starting fire is depicted on Figure 4.

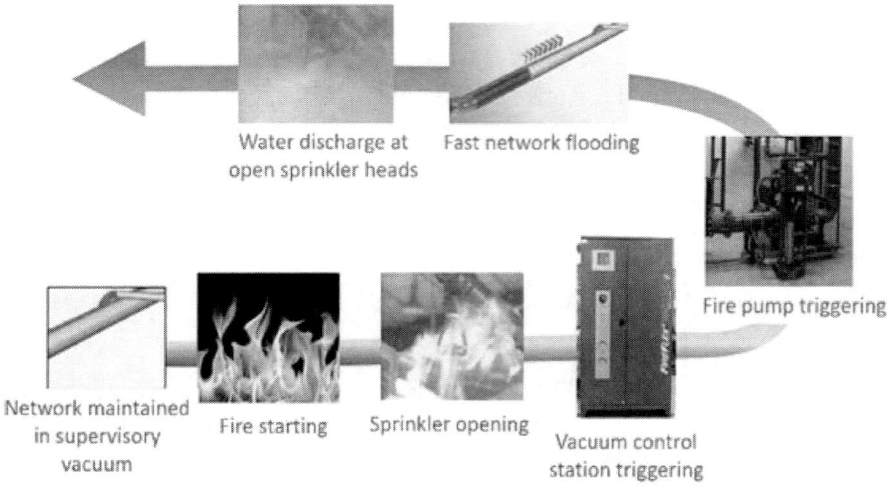

Figure 4. Illustration of the principle of a vacuum sprinkler system.

However, it is important to evaluate the performances of this system as compared to existing systems and to the applicable fire safety regulations, in particular regarding the time of arrival of water to the most disadvantaged sprinkler following its opening, the trip of the fire pump, and the evacuation of the air in the pipes for these specific pressure conditions. Thus, the aim of the present work is to provide a detailed study of the performance of dry-pipe and vacuum sprinkler systems on a pipework of Gridded typology. For this purpose, Computational Fluids Dynamics simulations were carried out, as detailed in the following section. The risk and location of air pockets trapped in the piping depending on the initial pressure is assessed as well for two different initial pressure values (+2.5 bar and -0.18 bar relative pressure). The latter is in fact an important information in particular for sprinkler systems of gridded typology which are usually not considered appropriate in the standards for use in dry pipe conditions, because of the occurence of this phenomenon. Furthermore, scale 1 test bench measurements were carried out in order to validate the simulation results obtained using the developed CFD model.

Material and Methods

Case Study

In order to asses the performance of the innovative vacuum sprinkler system, a a typical sprinkler system of Gridded typology was considered (Figure 5). Gridded systems comprise a main and secondary header, which allows to reduce the weight on the building's structure – this can be an advantage for protecting historic buildings. The volume of the pipework is $3.6m^3$. It comprises 20 branches with a horizontal spacing of 3m for a total length of 58.5m, and a width of 20m. The main header has an internal diameter of 160mm and the side branches have an internal diameter of 55.7mm. The secondary header has an internal diameter of 108mm (4"). The K80 sprinkler head (i.e., 80 l/min/bar1/2) is not modelled in full detail, but as an orifice of 19mm diameter.

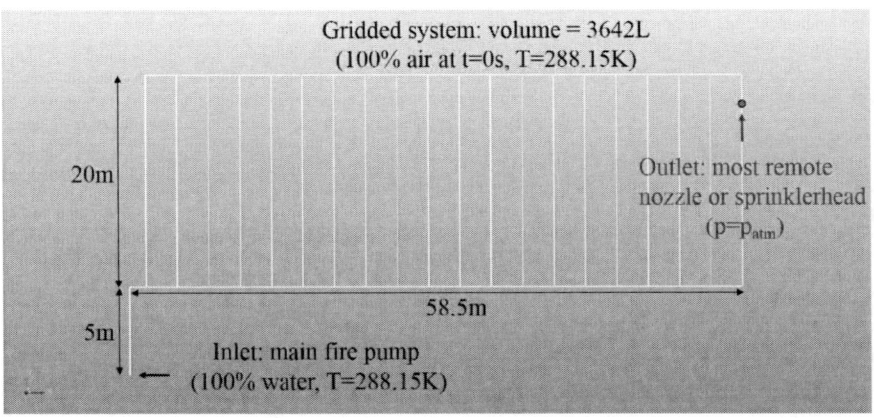

Figure 5. View of the Geometry of the Gridded typology.

In real systems, a dry pipe or vacuum valve is set to trip when a given pressure change is detected in the systems. In the simulations, the dry pipe/vacuum valve and the pump are considered to be at the same location, and the pump is set to start when the dry pipe valve trips. For all scenarios, it is assumed that only the most remote sprinkler head is opened (worst-case scenario). In the simulations, the sprinkler head is not modelled in full detail, but a specific routine is specified to calculate the outlet flow rate as a function of pressure and water fraction at the sprinkler head.

Experimental Method

The experimental setup, which corresponds to a simplified sprinkler system is depicted in Figure 6. The actual measurements are carried out only at the end of the whole setup comprising a main header/piping section of volume 1.168m^3 (internal diameter 102 mm, i.e., scale 1 in terms of pipe dimensions), and a tank of 2.5m^3 with mounted K80 (l/min/bar$^{1/2}$) sprinkler head. A water reservoir and a fire protection pump, deluge valve to control the release of water, as well as instrumentation are located upstream of the measuring section. The aim of the setup was to provide measurements on a simplified geometry of similar volume to validate the results from the CFD simulations. Macroscopic values, i.e., the activation time of the mechanical or electronic controller (resulting from the detection of a given pressure variation or pressure variation rate in the system), and the time for water to reach the sprinkler head after opening of the sprinkler head were compared.

The experiment begins with an initial total absolute pressure of 0.82 bar in the pipework (-0.18 bar relative pressure). At t = 0s, the sprinkler head is manually opened, resulting in a rise of the pressure in the system. When the pressure at the location of the main fire pump reaches 7 bar, the pump is activated and starts supplying water into the pipework according to the pump curve. The measurements are recorded all along the experiment. The experiment ends when a steady-state flow is reached at the sprinkler head, i.e., after approximately 6s. The repeatability of the test was ensured by performing the experiment twice.

The total pressure is measured by means of BALUFF pressure transducers in the range -1/10 bar relative pressure at an accuracy of +/- 2%. The measurements were carried out at a frequency of 100 Hz for all variables and recorded with the help of a DATAQ data acquisition system. The arrival of water was measured by means of a manual chronometer, with an assumed measurement uncertainty within +0.3 to +1s (resulting from the reaction time of the experimenter). The temperature in the room was measured with a type K thermocouple, and the regulation of the pressure in the system is ensured by the pressure maintenance pump and associated control system.

Figure 6. Experimental test bench. Left: Water reservoir, Scale 1 Fire Pump and pressure regulation unit. Right: Scale 1 pipework and water tank with K80 sprinkler head.

Numerical Method

Regarding the numerical method, the Navier-Stokes equations for a transient, 3D, turbulent, isothermal, two-phase flow were resolved over the discretized calculation field in the scope of the opensource finite volume CFD code OpenFOAM v3.0+ [23]. The *compressible InterFoam* solver was used, which consists in a Volume of fluid (VOF) approach to compute the interface between the gas (air) and liquid (water) phases. In this model, a single momentum equation is solved, and the momentum and other fluid properties are considered for the mixture. The corresponding equations (Eq. 1-4) are described below. In addition, the k-ε realizable turbulence model with wall function was used, leading to two additional transport equations for the turbulent kinetic energy k, and for the turbulent dissipation rate ε. Water was considered as incompressible (constant density), air as compressible perfect gas assumption and isothermal compression), and the PIMPLE algorithm was used for pressure-velocity coupling. Furthermore, adaptable time-steps were chosen such as to verify a CFL condition $C < 0.5$, ensuring the stability of the calculation, and leading to time-steps of 1e^{-5} to 1e^{-4} s.

Continuity Equation

$$\frac{\partial \rho}{\partial t} + \nabla.(\rho U) = 0; \text{ with } \rho = \alpha_l \rho_l + \alpha_v \rho_v \text{and } U = \alpha_l U_l + \alpha_v U_v \quad (1)$$

with U: mixture velocity, ρ: mixture density, l: liquid phase, v: vapor phase

Momentum Equation

$$\frac{\partial \rho U}{\partial t} + \nabla.(\rho \mathrm{UU}) = -\nabla(P - \rho \mathrm{gh}) - \mathrm{gh}\nabla\rho + \nabla\tau + \rho g + F_S \quad (2)$$

with g: gravitational constant, P: total pressure, τ: friction force, Fs: surface tension force

Energy Equation

$$\frac{\partial \rho T}{\partial t} + \nabla.(\rho \mathrm{UT}) = -\nabla(K\nabla T) + \left[\nabla(\mathrm{PU}) + \frac{\partial \rho K}{\partial t} + \nabla(\rho \mathrm{UK})\right]\left(\frac{\alpha_l}{C_{\mathrm{vl}}} + \frac{\alpha_v}{C_{\mathrm{vv}}}\right) = 0 \quad (3)$$

with T: temperature, K: diffusivity

Phase Continuity Equation

$$\frac{\partial \alpha_l}{\partial t} + \nabla.(\alpha_l U) + \nabla(\alpha_l \alpha_v U_r) = \alpha_l \alpha_v \left(\frac{\psi_l}{\rho_l} - \frac{\psi_v}{\rho_v}\right)\frac{\mathrm{DP}}{\mathrm{Dt}};$$

$$\text{with } U_r = U_l - U_v \quad (4)$$

with ψ: compressibility of the medium

Specific routines were then developed in order to take into account: (*i*) the main fire pump and (*ii*) the air and water flow discharged through the opened sprinkler head. A pump performance curve was specified at the Inlet of the computational field, coupling the total pressure to the water flow rate, with an assumed turbulence intensity of 5%. The outlet velocity conditions were calculated depending on the total pressure and water fraction at the outlet upstream of the sprinkler head according to Eq. (5-9) based on the Saint-Venant and Wantzel equations (Balachandran, 2006) and taking into account the sprinklerheads characteristics (K-factor, outlet area). The *choked flow* phenomenon, i.e., capped volume flow rate when sonic velocity is reached, was as well taken into account (Eq. 7). Furthermore, considering the large dimension of the calculation domain and resulting computational cost, the choice was made to consider a simplified geometry, i.e., a 2D grid extruded by 1 cell in the Z (vertical) direction with a symmetry boundary condition. This approach is equivalent to considering a simplified geometry. The adequacy of the 2D assumption was verified by comparing the numerical predictions to measurements on a scale 1 experimental setup in controlled conditions.

If $\alpha = 0$ (100% air)

$$p < 101325: U = \sqrt{\frac{2\gamma}{\gamma-1} \mathrm{RT}(1-\frac{p}{p_{\mathrm{atm}}})^{\frac{\gamma-1}{\gamma}}} \quad (5)$$

$$101325 < p < 191801: U = \sqrt{\frac{2\gamma}{\gamma-1} \mathrm{RT}(1-\frac{p_{\mathrm{atm}}}{p})^{\frac{\gamma-1}{\gamma}}} \quad (6)$$

$$p > 191801: U = 340 \ \textit{(choked flow)} \quad (7)$$

If $\alpha = 1$ (100% water)

$$U = \frac{K}{S}\sqrt{p - p_{\mathrm{atm}}} \quad (8)$$

If $0<\alpha< 1$ (air and water mixture)

$$U = \alpha\frac{K}{S}\sqrt{p - p_{\mathrm{atm}}} + (1-\alpha)\sqrt{\frac{2\gamma}{\gamma-1} \mathrm{RT}(1-\frac{p_{\mathrm{atm}}}{p})^{\frac{\gamma-1}{\gamma}}} \quad (9)$$

With α : water mass fraction (-), U : velocity magnitude (m/s), p : absolute pressure (Pa), p_{atm} (atmospheric pressure = 101325 Pa), T : fluid temperature (K), γ :adiabatic index, R: ideal gas constant (J/mol/K), K : discharge coefficient of the sprinkler (l/min/bar$^{1/2}$), and S: orifice section of the sprinkler head (m^2).

RESULTS

Grid Sensitivity

A grid sensitivity study was first carried out in order to choose the grid density for the computational domain, and to ensure independence of the results to the grid size. The simulations were carried out in for one case (pressure= -0.180 bar rel.) with electronic accelerator. The obtained results in terms of physical macroscopic values are presented in Table 1. The grid size G2 (8e-4m size - in the logarithmic layer - at 0.015m) provided the best tradeoff in terms of accuracy of results/calculation time and was retained for the rest of the study. Furthermore, the uncertainty in terms of water delivery time due to the grid will be assumed to be lower than 1s for the rest of the study.

Table 1. Grid sensitivity study

Grid density	Grid size	Activation time	Water delivery time	Calculation time (CPU)
G1	226k	2,27 s	20,80 s	~ 576 h
G2	296k	2,33 s	22,05 s	~ 960 h
G3	388k	2,34 s	22,85 s	~ 1728 h

The CFD model was first applied to a sprinkler system of Gridded typology, for which the use in dry-pipe conditions is typically not allowed because of potential instability due to trapped air pockets in the system. Simulations were hence carried out to further study this phenomenon and assess the location of the air pockets on a Gridded system comprising 20 branches, and with a system volume of 3.642m^3.

Experimental Validation

Measurements on the test bench presented in the above section, representative of a simplified sprinkler network, and of volume similar to the case studied numerically were then carried out. Two technical variants were considered for the triggering conditions, i.e., with mechanical accelerator (Mech.) and with electronic accelerator (Elec). The results presented in Table 2 show a very good prediction of the activation time ($\Delta t < 1s$) for all the cases, and a good to reasonable prediction of the arrival time of the first drop of water to the sprinkler ($\Delta t < 1s$ for the Mech. case and $\Delta t = 4s$ for the Elec. case). A water arrival criterion of less than 60s after detection of the fire is generally required in practice. Hence, the difference of 4s obtained for the Elec. case can be considered as acceptable, validating the assumptions made for the development of the CFD model.

Table 2. Comparaison of the experimental and numerical results for a relative pressure of -0,180 bar rel. and a K80 (l/min/bar$^{1/2}$) sprinkler

	Measurements (Mech./Elec.)	CFD (Mech./Elec.)
System volume (m^3)	3,668	3,642
Activation time (s)	11,5/2,0	11,36/2,33
Water delivery time (s)	33/26	/22

Single-Interlock Simulations

For the single-interlock case, the pump is activated at the beginning of the simulation (t = 0s), for instance consequently to the activation of a smoke detector. The pump supplies water into the piping, but the sprinkler head remains closed until its temperature rating is reached. This system allows to prevent false activations due to sensor defaults, which is particularly important for cultural heritage buildings. For these simulations, two different initial pressure values in the system were considered, i.e., 0.18 bar rel. corresponding to the vacuum system (also referred to as Case 0) and +2.5 bar rel. corresponding to the dry-pipe system (referred to as Case 1). The obtained results are presented in Figures 7-12 in terms of pump curve, pump pressure, water volume in the system, and water fraction distribution. For this scenario, the water flows in through the pump, and the pressure gradually increases in the system, with dynamic pressure gradually being converted into static pressure, see Figure 8. Finally, the system pressure reaches the pump churn pressure (zero flow). A dynamic equilibrium is obtained and the compressibility of air trapped in the system is responsible for a wave-like motion and redistribution of pressure. On Figure 8, it can be seen that the pressure at pump level remains low for a longer time for Case 0 (pressure between 5.2 and 6.5bar until t = 10s) than for Case 1 (p = 8bar at t = 10s). This results from the fact that the pump is facing a lower initial pressure in Case 0 than in Case 1. As a consequence, a higher water flow rate is delivered by the pump into the system (Figure 9), as it can be understood by the pump curve on Figure 7 – 15000l/min at 5.2bar versus 10000l/min at 8bar. As a consequence, a faster filling occurs for the vacuum system than for the dry pipe system, with 90% of water in the system at t = 20s, vs 57%, respectively (Figure 10). A better filling is observed as well, with 96% of the system filled with water at equilibrium for Case 0 vs 67% for Case 1, which results from the fact that a lower quantity of air is present for -0.18 bar rel. than for +2.5 bar rel. (perfect gas assumption). The resulting water fraction in the system is shown on Figure 11 and 12. It can be observed that a large air pocket is present for the dry-pipe system (initial pressure of +2.5 bar rel.) which could hinder the proper discharge of water if a sprinkler head was to open itself at this location, while the air pocket is significantly reduced for the vacuum system (initial pressure lowered to -0.18 bar rel.), resulting in a more stable system.

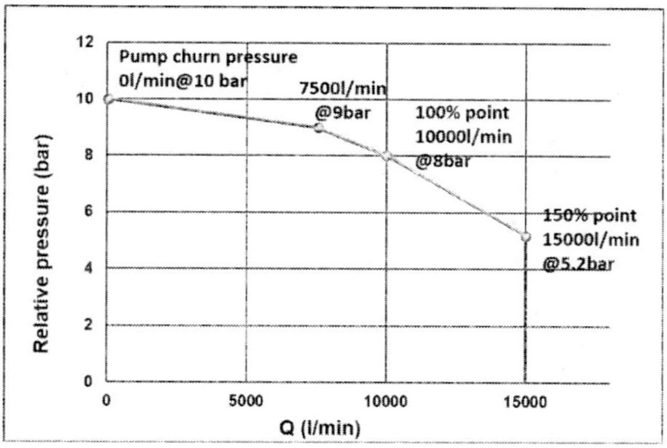

Figure 7. Pump characteristic curve.

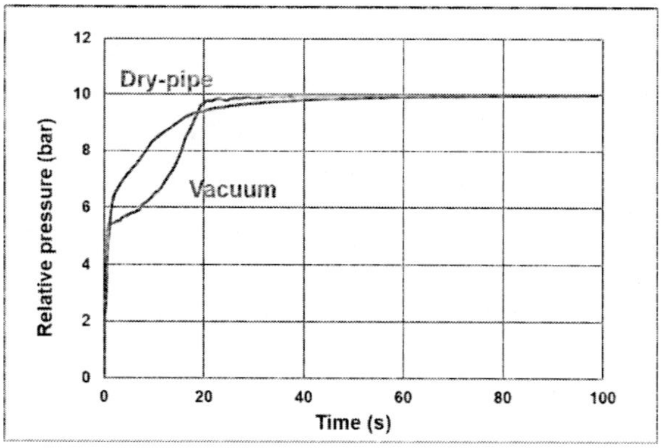

Figure 8. Relative pressure downstream of the pump after valve activation.

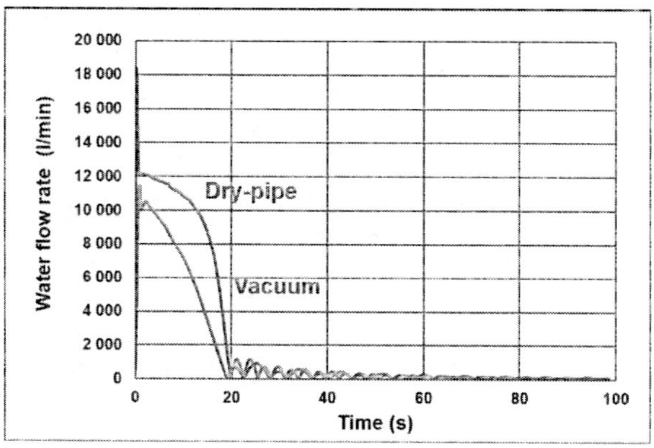

Figure 9. Inlet water flow rate.

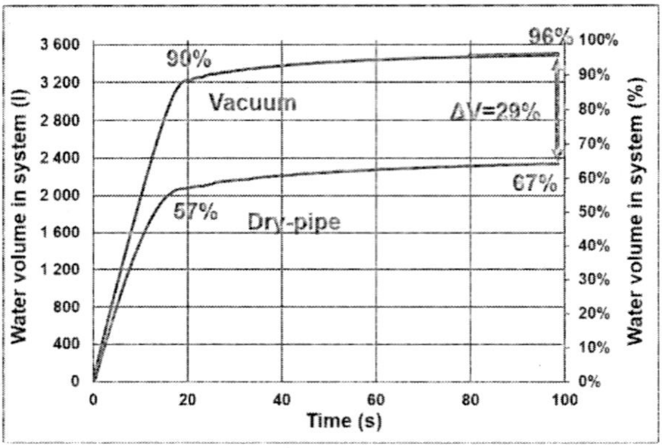

Figure 10. Water volume in system.

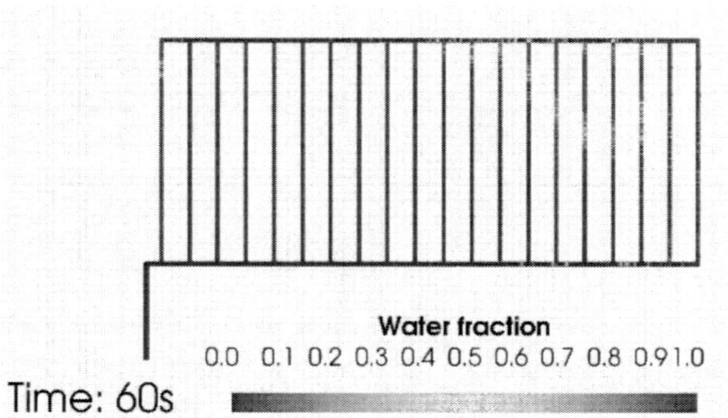

Figure 11. Single interlock: View of the water fraction 60s after valve activation for an initial pressure of -0.18 bar rel. (vacuum).

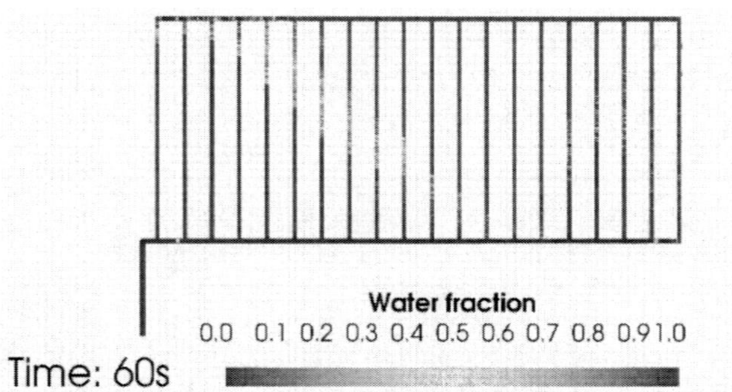

Figure 12. Single interlock: View of the water fraction 60s after valve activation for an initial pressure of +2.5 bar rel. (dry-pipe).

Double-Interlock Simulations

The second scenario concerns a double-interlock system: for this configuration, a sprinkler head is opened, but the pump is only activated after a delay which depends on the detection of a pressure variation Δp in the system subsequently to the sprinkler opening. Different technologies are considered which can be installed depending on the considered risk level: without accelerator (Δp = 1.38bar), mechanical accelerator (Δp/Δt = 14000Pa/s also referred to as VORV) and electronic accelerator (Δp/Δt = 700Pa/s also referred to as FLXPC).

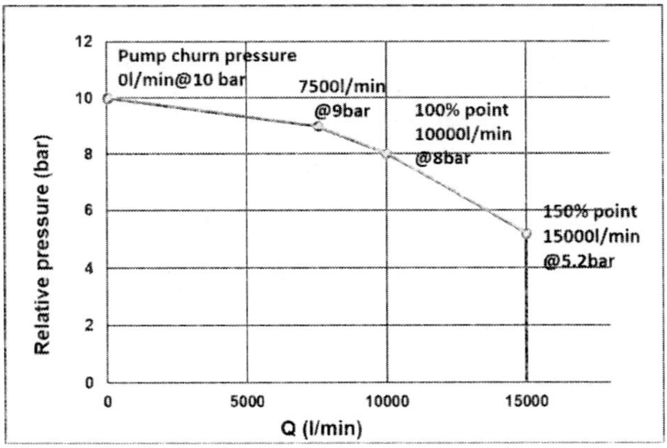

Figure 13. Pump characteristic curve.

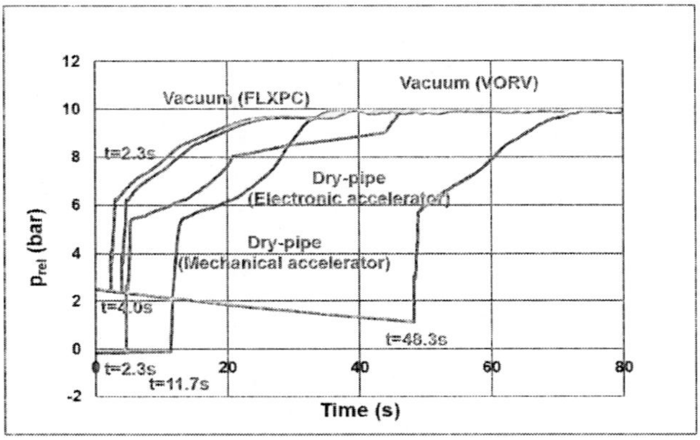

Figure 14. System pressure.

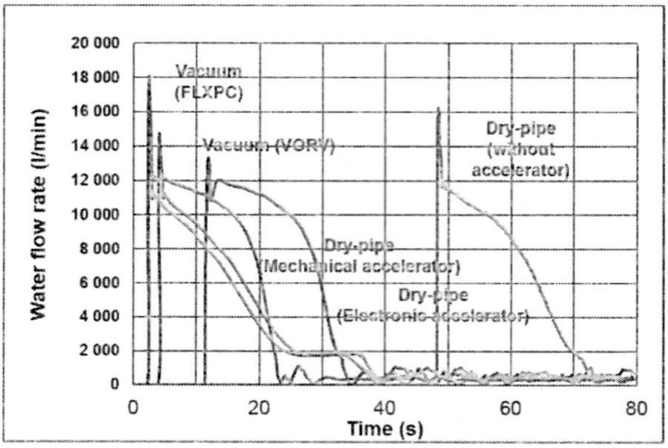

Figure 15. Inlet water flow rate.

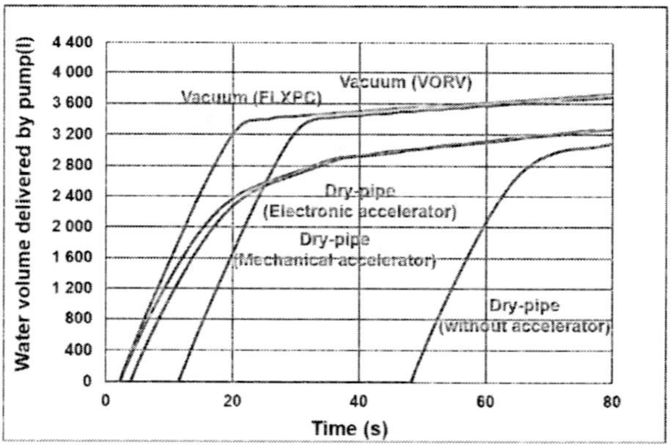

Figure 16. Total water volume delivered by pump.

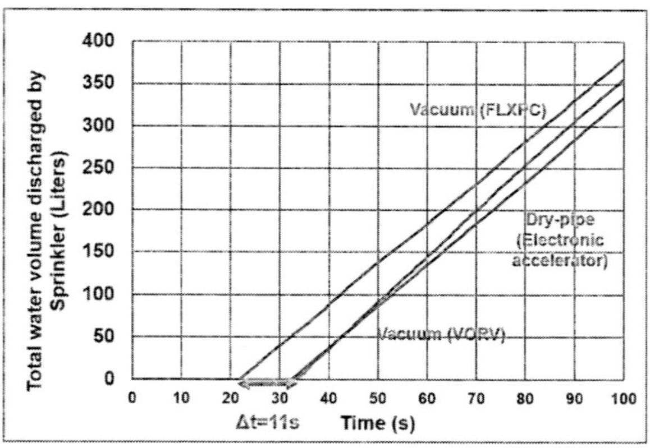

Figure 17. Total water volume discharged by sprinkler.

The results are presented on Figures 13-19 in terms of pump curve, inlet pressure, inlet water flow rate, and total water volume delivered by the pump and by the sprinkler. In this case, a competition of two phenomena occurs, on the one hand the discharge of air and water through the open sprinkler, and on the other hand the compression of air caused by the moving water front. Similarly to the previous scenario, a lower initial pressure in the system results in the pump delivering a higher water flow rate for a longer time, see Figure 14. As a consequence, a steady state water flow is reached at the sprinkler quicker for the vacuum cases than for the dry-pipe cases, with 23s to 35s, to a least 38s to 75s necessary, respectively depending on the technology used (constant slope on Figure 17). The water fraction distribution is as well presented on Figure 18-19 for this double-interlock scenario. For p_{init} = +2.5 bar rel. (dry-pipe), although a fraction of the air initially present in the system can be discharged through the open sprinkler head, the water follows the path of least resistance and trapped air pockets are blocked in the branches. For the -0.18 bar rel. case (vacuum), the pressure in the system is closer to atmospheric pressure at valve activation. As a consequence, less air is present in the system to be discharged, and a more homogeneous distribution of water is obtained at steady state, leading to an increased stability, see Figure 18 as opposed to Figure 19.

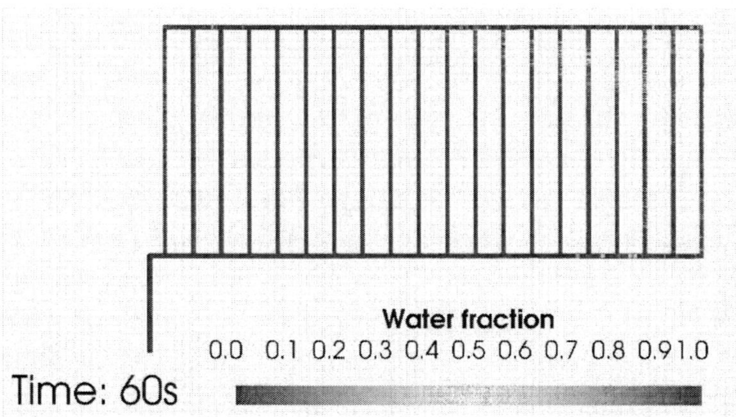

Figure 18. Double interlock: View of the water fraction 60s after valve activation for an initial pressure of -0.18 bar rel. (vacuum).

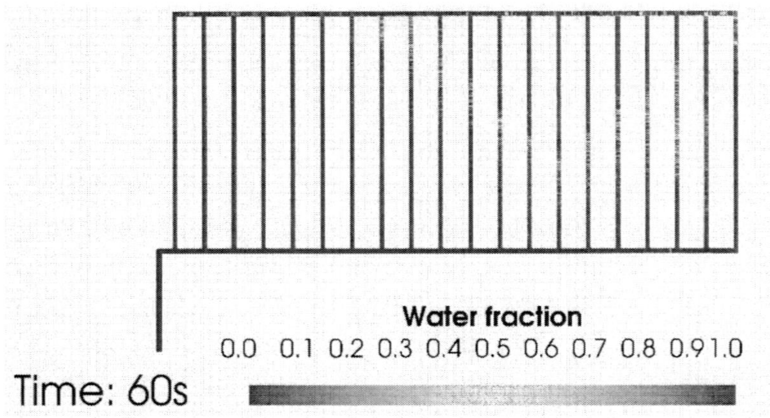

Figure 19. Double interlock: View of the water fraction 60s after valve activation for an initial pressure of +2.5 bar rel. (dry-pipe).

CONCLUSION

In this chapter, the specific requirements of historic buildings (e.g., cathedrals, castles, museums…) in terms of fire protection were introduced. Heritage buildings require tailored solutions for their automatic fire protection systems. The challenge is to provide an appropriate fire protection system with minimum impact on the building, and which is also resistant to corrosion, freezing, leaks or false activations. An innovative vacuum sprinkler system, working at a negative relative pressure (-0.18bar) was introduced. The performance of this system was evaluated by means of CFD simulations and measurements on scale 1 bench test for various scenarios. Specific routines were implemented in the framework of the CFD package OpenFOAM in order to take into account the fire pump, sprinkler head, and choked flow phenomenon. The compressible, turbulent, two-phase (air/water) flow was then solved inside of a sprinkler pipework of Gridded typology. The obtained results indicate that a lower initial pressure in the system could allow to reduce the water delivery time by up to 40%, and to increase the stability of sprinkler systems of gridded typology by reducing the phenomenon of blocked air pockets in the pipes. Preliminary experiments on a scale 1 experimental setup have validated the simulations results on a simplified geometry, both in terms of activation time and water delivery time. These results indicate improved fire safety performances for the protection of cultural heritage buildings equipped with such sprinkler systems.

ACKNOWLEDGMENTS

The authors gratefully acknowledge the support from the Centre-Val de Loire Region and the French National Agency for Research (ANR) in the scope of the APR-IR EPIC and ANR LabCom GreenSprink projects.

REFERENCES

[1] Escobar, H. (2018). *In a 'foretold tragedy,' fire consumes Brazil museum.*

[2] Ferreira, T. M. (2019). Notre Dame Cathedral: Another case in a growing list of heritage landmarks destroyed by fire. *Fire*, 2(2), 20.

[3] Devi, K. S., & Sharma, T. D. (2019). Innovations in conservation of heritage museums and libraries from fire hazards. *In AIP Conference Proceedings* (Vol. 2158, No. 1, p. 020005). AIP Publishing LLC.

[4] Batiot, B., Boulet, P., Cablé, A., Chetehouna, K., Coppalle, A., Collin, A., Garo, J-P., Morvan, D., Richard, F., Rogaume, T. & Santoni, P. A (2019). Mechanisms, issues and scientific challenges implied in large scale fires. The case of Notre-Dame Cathedral of Paris. State of knowledge of French laboratories and research efforts to come. Report from the *"Fire" French National Research Group CNRS n°2864.*

[5] Manhhai (2019). "*The devastating Notre Dame Cathedral fire, in 19 photos*". Licensed under Creative Commons BY 2.0. To view a copy of this license, visit https://creativecommons.org/licenses/by/2.0/

[6] Milanez, F. (2018). "*Fire at the National Museum of Brazil, in Rio de Janeiro, on 2 September 2018*". Licensed under the Creative Commons Attribution-Share Alike 4.0 International license.

[7] Ahrens, M. (2017). US experience with sprinklers. *National Fire Protection Association*. Research, Data and Analytics Division.

[8] CFPA Europe (Apr., 2017). *Many fire protection systems are faulty.*

[9] American International Group (2013). Freeze-Up of Fire Protection Systems. *AIG Programs Loss Control Bulletin*.

[10] CFPA-E (2013). *CFPA-E Guideline No 30:2013 F* "Managing fire safety in historical buildings".

[11] Ibrahim, M. N., Abdul-Hamid, K., Ibrahim, M. S., Mohd-Din, A., Yunus, R. M., & Yahya, M. R. (2011). The development of fire risk assessment method for heritage building. *Procedia Engineering*, 20, 317-324.

[12] Lostetter, M. O., & Breunese, A. (2005). Fire safety aspects in cultural heritage a case study in historical Delft. *Heron*, 50(4), 321-340.

[13] Chien, S. W., Lien, C. C., Sie, H. R., & Song, Y. T. (2012). Disaster risk assessment methods and response plans for cultural heritage in Taiwan. *Collections*, 8(4), 331-347.

[14] Kidd, S. (2003). Risk improvement in historic and heritage buildings. *In the Colvin Trust-Uppark Seminar*.

[15] Park, S. G., Mishima, N., Noh, H. W., Yoo, J. S., Oh, S. H., Min, B. W., & Oh, Y. S. (2015). Countermeasure against Fire Disaster in Regional Heritage Villages on the Concept of ICT-Based Disaster Prevention Design. *International Journal of Contents*, 11(1), 62-68.

[16] Torero, J. L. (2019). Fire Safety of Historical Buildings: Principles and Methodological Approach. *International Journal of Architectural Heritage*, 13(7), 926-940.

[17] Pau, D., Duncan, C., & Fleischmann, C. (2019). Performance-Based Fire Engineering Design of a Heritage Building: McDougall House Case Study. *Safety*, 5(3), 45.

[18] Chorlton, B., & Gales, J. (2020). Fire performance of heritage and contemporary timber encapsulation materials. *Journal of Building Engineering*, 29, 101181.

[19] Chorlton, B., & Gales, J. (2019). Fire performance of cultural heritage and contemporary timbers. *Engineering Structures*, 201, 109739.

[20] Son, D. W., & Han, G. S. (2014). Evaluation Methods of Flame Retardants for Wooden Cultural Properties. *Journal of the Korean Wood Science and Technology*, 42(5), 590-596.

[21] Benfer, M., & Williams, E. (2018). Assessing the impact of fire extinguisher agents on cultural resource materials. *Fire Technology*, 54(1), 289-311.

[22] Jensen, G. (2006). Hypoxic air venting for protection of heritage. *COST-Action 17 Built Heritage: Fire Loss to Historic Buildings*. Riksantikvaren, Historic Scotland.

[23] Jasak, H., Jemcov, A., & Tukovic, Z. (2007, September). OpenFOAM: A C++ library for complex physics simulations. In *International workshop on coupled methods in numerical dynamics (*1000), 1-20. IUC Dubrovnik Croatia.

BIOGRAPHICAL SKETCH

Khaled Chetehouna

Affiliation: INSA Centre Val de Loire (National Institute of Applied Sciences), University of Orléans, PRISME EA 4229 Laboratory, France

Education: Hab.Dir.R., Ph.D., M.Eng.Sci.

Business Address: 88 boulevard Lahitolle CS 60013 18022 Bourges, France

Research and Professional Experience:

Dr. Khaled Chetehouna is Full Professor in Fire and Energy Sciences, and Vice-Chairman for Research at the National Institute of Applied Sciences (INSA) of the Centre Val de Loire region, France. He is head of the P2CF (Permeation Pyrolysis Combustion and Fire) team at the

PRISME Laboratory (EA 4229). His research activities are devoted to Fire Sciences and Fire Safety, and in particular: the characterization of Fire Degradation of Composites Materials for Aerospace and Automotive applications, the study of Flame and Smoke propagation in buildings, Forest fires, and the Evaluation of the performance of Fire Suppression systems for the protection of Industrial buildings and Cultural Heritage, including innovative environmentally friendly sprinkler systems. He is a member of the strategic board of the CNRS French Research Group in Fire Sciences GDR "Feux" CNRS n°2864 since 2010, and in charge of relations with Morrocco (ENSA Agadir and ENSA Khouribga), India (VelTech University) and Indonesia (BPPT : Agency For The Assessment And Application Of Technology) for the National Institute of Applied Sciences since 2015. His publications include more than 80 refereed articles with peer review in the fields of Fire Sciences, Engineering and Composites Materials.

Professional Appointments:

- Vice-Chairman for Research at the National Institute of Applied Sciences (INSA Centre Val de Loire)
- Full Professor at the National Institute of Applied Sciences (INSA Centre Val de Loire)
- Member of the strategic board of the CNRS French Research Group in Fire Sciences GDR "Feux" CNRS n°2864

Honors:

Holder of the French Grant for Scientific Excellence (2011-present)

Selected Publications from the Last 3 Years:

1. Rashid, M.,Chetehouna, K., Cablé A & Gascoin, N. (2020). Analysing Flammability Characteristics of Green Biocomposites: An Overview. *Fire Technology. (In press)*
2. Manescau, B., Chetehouna, K., Sellami, I., Nait-Said, R., & Zidani, F. (2020). BLEVE Fireball Effects in a Gas Industry: A Numerical

Modeling Applied to the Case of an Algeria Gas Industry. In *Fire Safety and Protection*. IntechOpen.

3. Manescau, B., Chetehouna, K., Serra, Q., Wang, A., & Florentin, E. (2020). Numerical Study on the Outdoor Wind Effects on Movement Smoke along a Corridor. In *Fire Safety and Protection*. IntechOpen.
4. Chetehouna, K. (2020, April). Overview of The Forest Fire Research. In Proceeding *International Conference on Science and Engineering* (Vol. 3, pp. xvii-xvii).
5. Dugast, G., Settar, A., Chetehouna, K., Gascoin, N., Marceau, J. L., Bouchez, M., & De Bats, M. (2020). Experimental and numerical analysis on the thermal degradation of reinforced silicone-based composites: Effect of carbon fibres and silicon carbide powder contents. *Thermochimica Acta*, 686, 178563.
6. Abed, S. A., Settar, A., Chetehouna, K., Kadiri, M. S., El Tabach, E., & Gascoin, N. (2020). Numerical study on a porous material subject to SiC particles deposition, using OpenFOAM and sensitivity analysis technique: Effect of clogging evolution on the thermal performances. *Chemical Engineering Science*, 212, 115321.
7. Cablé, A., Chetehouna, K., Gascoin, N., Settar, A., Van Herpe, K., & Kadoche, M. (2019). Experimental and numerical study of sprinkler fire protection systems: Effect of initial pressure on system performance. *Journal of Loss Prevention in the Process Industries*, 60, 76-86.
8. Wang, A. J., Manescau, B., Serra, Q., Chetehouna, K., Florentin, E., & De Izarra, C. (2019). Numerical simulations of outdoor wind effects on smoke spreading along a corridor: Physical and sensitivity analysis. *International Journal of Thermal Sciences*, 142, 332-347.
9. Coudour, B., Chetehouna, K., Lemée, L., Bertin, P., & Garo, J. P. (2019). Thermal degradation of α-pinene using a Py–GC/MS. *Journal of Thermal Analysis and Calorimetry*, 137(4), 1315-1328.
10. Grange, N., Manescau, B., Chetehouna, K., Gascoin, N., Lamoot, L., Coppalle, A., & Reynaud, I. (2019). Fire resistance of carbon-

based composite materials under both ideal and realistic normative configurations. *Applied Thermal Engineering*, 159, 113834.

11. Taddeo, L., Gascoin, N., Chetehouna, K., Ingenito, A., Stella, F., Bouchez, M., & Le Naour, B. (2019). Experimental study of pyrolysis-combustion coupling in a regeneratively cooled combustor: Heat transfer and coke formation. *Fuel*, 239, 1091-1101.
12. Najmi, H., El-Tabach, E., Gascoin, N., Chetehouna, K., & Falempin, F. (2019). Axial Distribution of Permeance and Selectivity of a Porous Cylindrical Tube for Binary Gas Mixtures (CO2/N2). *Industrial & Engineering Chemistry Research*, 58(8), 3359-3369.
13. Saadallah, S., Cablé, A., Hamamda, S., Chetehouna, K., Sahli, M., Boubertakh, A., & Gascoin, N. (2018). Structural and thermal characterization of multiwall carbon nanotubes (MWCNTs)/ aluminum (Al) nanocomposites. *Composites Part B: Engineering*, 151, 232-236.
14. Sellami, I., Manescau, B., Chetehouna, K., de Izarra, C., Nait-Said, R., & Zidani, F. (2018). BLEVE fireball modeling using Fire Dynamics Simulator (FDS) in an Algerian gas industry. *Journal of Loss Prevention in the Process Industries*, 54, 69-84.
15. Chetehouna, K., Grange, N., Gascoin, N., Lemée, L., Reynaud, I., & Senave, S. (2018). Release and flammability evaluation of pyrolysis gases from carbon-based composite materials undergoing fire conditions. *Journal of analytical and applied pyrolysis*, 134, 136-142.
16. Grange, N., Chetehouna, K., Gascoin, N., Coppalle, A., Reynaud, I., & Senave, S. (2018). One-dimensional pyrolysis of carbon based composite materials using FireFOAM. *Fire Safety Journal*, 97, 66-75.
17. Grange, N., Tadini, P., Chetehouna, K., Gascoin, N., Reynaud, I., & Senave, S. (2018). Determination of thermophysical properties for carbon-reinforced polymer-based composites up to 1000°C. *Thermochimica Acta*, 659, 157-165.

In: Critical Issues and Analysis … ISBN: 978-1-53618-738-0
Editors: A. Cablé and B. Manescau

Chapter 6

WILDFIRES PRE-ATTACK PLANS FOR THE URBAN INTERFACE AREA, AS A FIRST RESPONSE STRATEGIES – VALPARAÍSO-CHILE

Miguel E. Castillo*

Wildfire Laboratory, University of Chile,
Santiago, Metropolitan Region, Chile

ABSTRACT

The response time for the attention of an emergency caused by a forest fire in inhabited areas is the determining factor to evaluate the possibilities of control against the advance of the fire. This time depends on factors related to vegetation, the construction material of the houses, the local infrastructure and weather conditions. However, a little studied factor is the division and qualification of the territorial units in priorities for emergency care. This chapter shows a case study based on the creation of a set of criteria aimed at establishing a mapping classified as first attack priorities, based on a multi-criteria analysis. The Valparaíso Commune (Chile) was considered for presenting an extensive history of serious and

* Corresponding Author's Email: migcasti@uchile.cl.

extensive fires at the urban-forest interface. This method is fully applicable to other inhabited areas that present a serious danger to the occurrence of forest fires near homes.

Keywords: wildland urban interface, wildfires, response times, fire behaviour

INTRODUCTION

Forest fires have traditionally been a major problem in most regions of the earth, due to the serious losses that fire causes in renewable natural resources and the severe environmental and social impacts that are generated. The damages, especially in recent times, are also affecting the housing sectors located at the urban-rural interface, with the destruction of homes and the impacts on people themselves, with repercussions and disasters of an amount difficult to imagine.

In the year 2020, for example, 8,125 forest fires have been registered in Chile according to the count obtained until the month of June of the year in question, which is equivalent to an 8.2% increase over the existing average for the period 2015-2019 (5 years). Similarly, this season there was direct damage of 101,577 hectares, which corresponds to a value higher than the annual average in Chile.

The scenario just described and its projections are clearly adjusted to the situation of the Communes of Valparaíso and Viña del Mar, located on the coast of the V Region of Chile. This area is characterized by a high occurrence of forest fires in densely populated sectors, and where, year after year, the media report on the countless homes devastated by fire and the unfortunate damage to its inhabitants, including the loss of lives.

Currently, as a result of climate change, and the evolution of social scenarios, conditions are being created that increase the occurrence and intensity of forest fires and, consequently, the severity of the ecological and socioeconomic damages and impacts that they produce. In this sense, the possibility of having instruments with greater capacity and efficiency to evaluate the losses and effects of fire in the natural environment, constitute a valuable opportunity to have better foundations for its prevention and fire fighting.

Traditionally, the allocation of resources for forest protection has been determined by first-approximation indicators, easy to use, but of little scientific depth. Now, the greater availability of mathematical tools and algorithms allows the incorporation of new variables in the analyses and, also, facilitates the development of better models to support decisions for forest defense [1-3].

On the other hand, the real capacity is established to include factors associated with the economic valuation of natural resources, the net change in their value as a result of the impact of fires, operational performance and the costs of extinguishing media, to mention some of the important possibilities that are opened thanks to technological development [4, 5].

THE INTERFACE AND WILDLAND FIRES

With reference to interface zones in the Valparaíso region, Castillo [6] explains that one of the most complicated problems in developing a fire is the speed at which the fire progresses through the hills and houses. Reference values taken from the real evidence indicate average speeds in a range of 40 to 300 m^2min^{-1}. However, in extreme fire conditions, this speed has been seen to exceed 600 m^2min^{-1} in some cases, which is equivalent to more than 3.2 ha hr^{-1} [7]. An even more extreme scenario occurred on 2014 in Valparaíso. According to air and field monitoring systems, this immense fire advanced at speeds of up to 1,300 m^2min^{-1}, equivalent to approximately 8 ha hr^{-1}, unprecedented figures that are outside any measurement scale.

Regarding the construction material of houses at risk of fire, the behavior of fire in structures and buildings is apparently different from that of open areas with fuel vegetation [8]. In these areas of ignition, the rate of fire spread and the heat quantities generated by the combustion of lightweight structural materials have been studied using mathematical models [9] that show that expansion can generate columns of fire that favor the formation of hot air cyclones along with incandescent gases within ventilated structures. Pyne et al. [10] reported flame residence times much higher than those generally occurring in the combustion of arboreal or shrub vegetation in open spaces. This ensures that the possibility of rapid fire spread through residential buildings is even greater if the density of combustible buildings is also high. In many cases, residents living in areas at high risk of forest fires prefer to accept the existence of fires and organize to fight the fires, rather than taking the necessary precautions to minimize the risk of fire ignition and spread.

The hazard of fire spread depends on different factors. The main factors are: wind (speed and direction), topography (terrain slope) and fuel vegetation. The first two factors are practically impossible to control in a pre-attack plan, while the vegetation can be controlled through proper forest management in the surrounding areas, ravines and densely populated centers, which undoubtedly reduces the danger of fire. This type of forest management involves a variety of activities such as pruning, reducing the amount of combustible material in horizontal and vertical strata, extensive cleaning of garbage dumps on the slopes, increasing thinning of vegetation and strengthening access roads for eventual fire control [11, 12]. However, in many cases these practices are not carried out and, in others, they are carried out poorly or to an insufficient degree in terms of the level of protection required, especially in those places where the presence of high slopes and winds Intense premises could trigger large fires Rapid spread [11]. To reduce and prevent this type of tragedy, nationally applied standards have been created in the United States to reduce WUI risk circumstances, regulated by the NFPA (National Fire Protection Association) and associated collaborative programs [13].

These standards consider the commitment of voluntary organizations, programmed community activities in risk areas, emergency protocols in relevant situations and state support through the activation of direct aid mechanisms [7, 14]. Faced with the seriousness of the forest fire problem at this interface, several educational projects have been carried out in the United States, coordinated by the NWCG (National Committee for Coordination of Forest Fires), through participatory workshops with local communities on the management of fuel vegetation, reaction strategies in emergency situations and collective behavior [15]. The goal of all of these measures is to reduce the fire hazard in the interface area. In Chile, some activities have been carried out at the local level and have not had the necessary continuity or adequate evaluation over time to be formally available as part of an integrated fire prevention program.

Frequently, it is possible to observe combinations or mixtures in this interface zone. Davis [16] and Radeloff et al. [17] distinguish five categories or combinations of situations in this zone or WUI interface. For example, the WUI band or area that is exactly that, made up of clearly delimited segments that separate closely linked structures and consolidated by other provisional or less developed segments; an intermixed area, composed of scattered structures but always maintaining the continuity of the existence of fuel. Also an occlusion area, made up of cohesive islands or fragments of combustible material, often small constructions made of lightweight construction material; in the past, there was a border area where there were populated settlements next to fuel enclaves of combustible vegetation in sectors with variable topography; and finally vegetation areas, in many cases around the aforementioned areas [18]. In the case of Valparaíso, Rodríguez et al. [19] carried out an exhaustive and detailed registry of homes in interface areas, determining four categories in terms of construction materials and their location on the slopes. The structural properties and particular characteristics of the locations of these homes allowed them to be associated with a fire hazard factor. As a result, a preliminary economic evaluation was obtained, as well as an estimate of the number of houses located at the interface that could be affected by future forest fires.

Closely related to the above is the issue of civil protection and emergency evacuation strategies. The first aspect to consider in a fire protection strategy in interface areas is the safety of the civilian [7]. The fundamental role of the state in the face of any type of catastrophe is to guarantee at least the basic mechanisms necessary for the evacuation, combat and distribution of the resources necessary to face the problem of fires in the interface areas, especially when the spread of fire acquires aggressive characteristics.

Typically, traditional mechanisms begin with geographic and financial analyzes of the resources available for combat, often neglecting prevention strategies such as the proper management of fuel vegetation. When the meteorological, topographic and structural conditions of the risk areas are favorable for the ignition of the fire, defined warning categories are applied, which involve different control processes that will be instigated when the fire is ongoing.

REFERENCES IN CHILE

Chile has a good statistical system for registering forest fires, operating from 1986 to the present year 2020. The evolution of the occurrence and damage has been studied throughout this period, showing significant changes in the recurrence of episodes critical since 1998, which is when the country presents with greater frequency and intensity prolonged episodes of droughts and consequently, an increase in the condition of danger. Figure 1 shows a synthesis of the number of fires for the last season (2018-2019), for each of the regions of Chile. A greater number of fires in the regions of the central territory stands out. Regarding the damage caused by forest fires in Chile, the figures are generally stable over time, except for the 2016-2017 season, when nearly 575,000 hectares were burned among forests, scrublands, grasslands and forest plantations. The variability of these figures has been notable, possibly due to the paucity of data prior to 1990, and also due to changes in the dynamics of occurrence and causality in the last 15 years. In the 2016-2017 season, the largest and most disastrous forest fires occurred in the contemporary history of Chile [20]. In just 3 months, thousands of hectares of native forest, scrub and plantations were razed by fire, further accentuating the need to develop initiatives and studies for the protection of wooded areas, and especially the infrastructure and inhabited sectors. In this area, the occurrence of urban-forest interface fires has been increasing, with the need for further research to improve emergency response and response times for the arrival of the first resources for wildfire fighting.

The meteorological factor has been a variable that has favored the uncontrolled spread of fire in inhabited areas. Chile has almost 15 years of accumulated drought, which only in 2020 has had a slight tendency to increase rainfall in the central zone. However, the condition of danger is maintained by the high density of homes located in wooded areas and with high potential for fire spread.

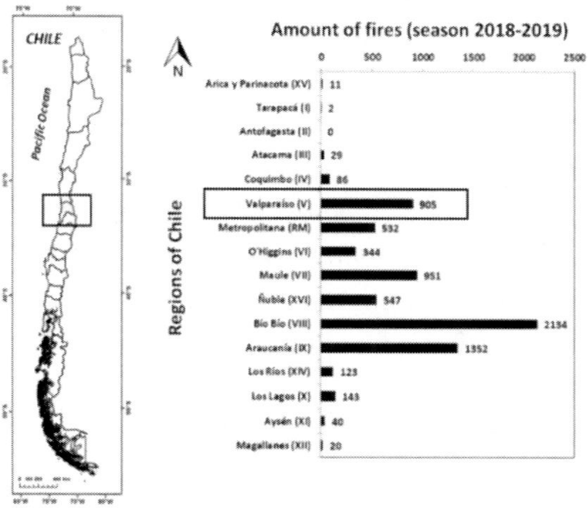

Figure 1. Distribution of wildfires in Chile in its territory.

PRE-ATTACK PLANS

Pre-attack plans are strategic actions previously prepared in order to optimize access to the most vulnerable areas in the face of the arrival of fire towards the houses and also as a preparatory act to combat forest fires. These plans must contain the type of mobilization to be carried out and even the combat modality to be applied in the control of a possible fire. This requires, therefore, a detailed and prior evaluation of the terrain, so that the fire simulation is reliable and, therefore, the formulation of strategies and the programming of the resources to be used.

Due to the need to carry out a detailed description of the sector on which the pre-attack is being formulated (topography, accesses, anchor lines, water sources, fuel models, local winds, availability and location of resources for fire fight), so as to support knowledge about the different possible fire suppression options, these plans should be reserved exclusively for sectors with a high priority of protection and with a high rate of occurrence.

In addition, it is necessary to bear in mind that the pre-attack plan must be reviewed by the fire chief, once the recognition of the origin of the fire on arrival at the place is completed, due to the impossibility of predicting with certainty the exact point at which it can start a fire and, therefore, the environmental conditions that may arise at the time of the event. For this reason, the pre-attack is intended to program mobilization and combat decisions in general terms, thereby achieving a gain of time that can be decisive in controlling the eventual fire.

The installation of pruning programs and cutting of adult and dense vegetation, in addition to the preventive role it plays, may represent an important component of the pre-attack, due to the sectorization established in the intervened area, in addition to various benefits that are achieved for the development of fire fighting.

The installation of penetration lines, as in the case of firewalls, should be carried out taking full advantage of topographical accidents, trails, waterways and other sectors without vegetation, in order to reduce their installation and maintenance costs. They should also be considered as an important component of pre-attack schemes.

In recognition, it is also appropriate to review and complete the decisions contained in the pre-attack plan, logically in the cases in which it has been formulated. In the pre-attack plan, although the strategy to be adopted in a certain place in the event of a forest fire occurs, it is not possible in most cases to define the specific actions (tactics) to follow, due to the natural uncertainty to accurately forecast the point of origin of the focus, the environmental conditions that will prevail in the opportunity and, also, the types and amounts of combat resources that will be available.

Study Area

The study was carried out in the Valparaíso region (Central Chile). It has an area of 16,396.1 km^2, representing 2.16% of the country's surface. It is located between 32º02' and 33º57' south latitude, and from 70º00' west longitude to the Pacific Ocean, including the Easter Islands, Salas and Gómez, San Félix, San Ambrosio and Juan Fernández Archipelago. It is administratively divided into 8 provinces and 38 communes. It is one of the regions that presents the largest number of forest fires in areas of urban-forest interface, with a history of serious fires that have destroyed thousands of homes and caused extensive damage to infrastructure and also to the neighboring forests. Figure 2 shows the main areas occupied by houses and other structures, which are located in areas of high vulnerability to fire.

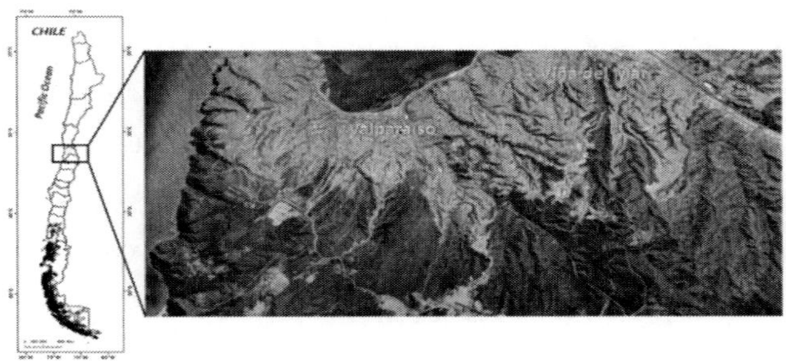

Figure 2. Satellite view of the metropolitan area of the cities of Valparaíso and Viña del Mar (Chile). In these areas, a high density of fires associated with homes and forests are concentrated in the urban forest interface.

Figure 3. The configuration of the city of Valparaíso has been carried out in the system of ravines that present steep slopes, formed from the degradation processes in the upper parts. People have been progressively using these areas for the construction of houses, which due to their location, are exposed to a high risk of fires. Courtesy of CONAF, 2015.

The destruction of houses in interface areas as a result of forest fires has been a constant problem with a high economic and social impact. Before 2001, the loss of houses in the Valparaíso region were isolated events, as was seen by the destruction of houses in "Population Gómez Carreño" in 1968 (200 houses) and in "Rodelillo" in 1994 (56 houses). However, since 2001, this problem has become increasingly serious with registered fires each year; The most dramatic cases were those of 2003 in "Cerro Cordillera" (34 houses) and "Miraflores Alto" (17 houses). In 2004, the damage was concentrated in "Villa Rapa Nui" (18 houses); 2008 in "Cerro La Cruz" (200 houses); 2012 in "Canal Beagle" (77 houses), 2013 "Cerro Rodelillo" (309 houses), "El Vergel" (30 houses) and in 2014 in "Camino La Pólvora" (2,900 houses). The latter has been classified as one of the most serious fires in the history of Valparaíso. In 2015, another large interface fire in Valparaíso with 553 hectares burned posed a serious threat to other populated areas that had not been affected in 2014. For the vast majority of these cases, the damage could have been reduced considerably, had there been a strategy modern and effective for civil protection and the timely response to actions of supression and combat. Later in 2019, another forest fire in December burned more than 300 homes. In 2020, hundreds of homes were again destroyed by fire in densely populated areas of Valparaíso. An illustration of the location conditions of these structures is shown in Figure 4.

Figure 4. Isolated houses and also forming groups. Example sector, in Valparaíso, Chile. Photo: Courtesy of CONAF, 2020.

In 2019 and 2020, different state institutions and also companies have been studying the interface defining firewall widths and areas free of dense vegetation. However, although standards currently exist, more study is needed on mathematical calculations to support, improve and maintain them.

METHODS AND RESULTS

The development of this case study begins with the cadastre of the areas with the highest occurrence of forest fires, considering the metropolitan area of Valparaíso. It is a territory that presents a high complexity due to the high density of houses in sectors with very difficult access, and a communication network in many areas, not prepared for the arrival of large vehicles and firefighters. Although there has been a significant improvement in the infrastructure for water supply and other basic services for fighting forest fires, there are still areas with a high risk of rapid fire spread because many houses are located in favor of the wind direction that the area predominates, and because the communities are also not necessarily prepared to carry out prevention and evacuation actions against the occurrence of emergencies. These are problems that have arisen in Valparaíso raise a priority issue for the institutions of the State of Chile because the magnitude of the damage has been repeated in recent years.

The process of formulating pre-attack plans begins first with a territorial analysis of the protection priorities for Valparaíso, based on criteria associated with the Risk, Danger and Potential Damage of forest fires; in parallel, all those areas associated with the Metropolitan Regulatory Plan and a general cadastre of the road infrastructure and support means for the free transit of vehicles associated with fire fighting were considered. This step was carried out by defining a scheme of variables and scores in a multi-criteria analysis, in order to establish areas with different categories of vulnerability. In parallel, a cadastre of fuel vegetation and a climatic and topographic characterization of the territory in question were carried out. The summary of this first stage is shown in Figure 5.

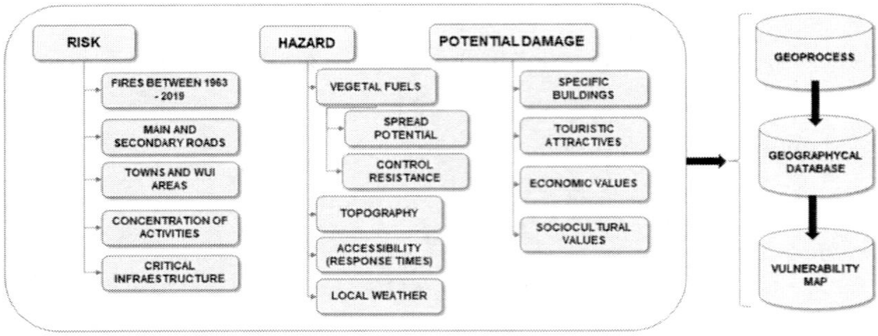

Figure 5. Variable scheme used for territorial analysis of protection priorities.

The allocation of scores for each of the variables was made through consultation sessions with experts, mainly people living in Valparaíso, also representatives of communities present in the areas of greatest vulnerability and who are also within sectors belonging to the Regulatory Plan at Metropolitan Zone of Valparaíso.

The determination of vulnerability for the different sectors of a protected area was based on the integration of the results of the analyses of three general variables: Risk (probability of fire initiation); Danger (intensity in the spread of possible fires that originate); and Potential Damage (losses and impacts that could be generated by the spread of fire). In turn, for each general variable it was necessary to identify the specific variables and subvariables that compose them, which must be defined by their relevance or influence on the corresponding risk, danger and potential harm present.

Along with the identification of the variables, it was necessary to define their weight or weight (normalized score on a scale from 0 to 1,000), in order to allow the quantitative assessment of each of the surface units considered in the application of the method. This assignment of normalized scores started with the general variables, keeping the precaution that the sum of the three upper limit values of their ranges was equal to 1,000. Subsequently, the value set for each of the general variables should be distributed among all the specific variables selected to compose it.

For the case of the study area, a panel of five participating experts was prepared with very extensive knowledge of Valparaíso and the problem of forest fires, and based on the feasible background of collecting and processing, the variables and subvariables were selected specific corresponding to each of the three general analyses, which are presented in Table 1.

Table 1. Scheme of variables and scores assigned in the process of determining protection priorities for the research area

	Specific criteria			
General criteria	Level 1	Level 2	Level 3	Score
Risk	Fire occurrence (1963-2019)			70
	Main and secondary roads			45
	Towns and WUI areas		Isolated houses	35
			House groups	50
	Concentration of activities			40
	Critical infrastructure		Basic services	50
			Other buidings	10
				300
Hazard	Vegetal fuels	Spread potential		55
		Control resistance		80
	Topography	Slope		100
		Exposure		15
	Accessibility			50
	Local Weather(*)	Winds		75
		Temperature		25
				400
Potential Damage	Specific buildings			70
	Touristic attractives			70
	Economic values			100
	Sociocultural values			60
				300

* The local climate is extremely relevant because in the study area, the main factor that determines the behavior of the fire is the wind (speed and direction), due to the proximity to the Pacific Ocean.

The score or weight of the selected variables (normalized scores) was defined by the aforementioned expert panel, based on the application of the Delphi Method. To this end, on the basis of an initial proposal of scores for the different variables, the group of experts through repeated discussion sessions on the various existing options concluded with a unanimous agreement on the final assignments. A basic agreement was given to grant an identical total weight for the set of variables related to the Risk and

Danger analyses with respect to those referred to the Potential Damage Analysis, corresponding to maximum or higher normalized scores of 300 + 200 and 500, respectively. Once the upper limits for the general variables were established, the discussion was continued to decide the assignments of the specific variables of the first level, to continue later with those of the second level and finally those of the third level. In this process, the established restriction was kept in mind that the sum of the scores of the specific variables of a certain level could not exceed the assignment of the highest level general or specific variable to which they were assigned. The final results of assigning the normalized scores can also be seen in Table 1.

In the case of Risk, 300 points were assigned (out of a total of 1,000 possible), which were organized into 5 specific criteria associated with the occurrence of fires (70 points), the infrastructure of main and secondary roads, especially those that are necessary for the entry and exit of the extinguishing media (45 points), the characterization and zoning of the areas at the interface (size, density, proximity to vegetation) (35 points) in which case the presence of isolated houses was evaluated and also groups of houses; also the concentration of activities associated with the mobility of people, vehicles and tourist traffic (40 points), and the condition of critical infrastructure in which the initiation of intentional fires in the vicinity of basic services and others has been verified associated with the proximity of the interface (60 points). The Hazard component was valued at 400 points, of which 135 were assigned to the characteristics of combustible vegetation (55 points for the potential for vegetation propagation and 80 points for resistance to control). A high amount of vegetation is present in the area that promotes the advance of the fire towards the upper parts of the topography. For this reason, a strong incidence of the slope was assigned (100 points). Regarding accessibility, in many locations it is complicated due to the poor condition or maintenance of entry and exit roads to the houses located at the interface (50 points); finally, in this specific criterion, there is the local effect of the wind, which acts in a very important way in the development and projections of the advance of the fire, and that in most emergencies, constitutes a decisive factor for the planning of combat strategies by air and land. For this reason, 50 weighting points were assigned. The effect of temperature is somewhat more transversal for the area under study (25 points), but it is included because at high values of this factor, it promotes the risk condition of starting and later of the spread of fires. Finally, the third general criterion considered in this study corresponds to the potential damage, described by the specific criteria associated with vulnerability in housing, in sectors of tourist activity, economic and social values (300 points). They are very relevant criteria due to the high magnitude of the damages that have occurred in previous years, and that currently there is no sign of a decrease in the condition of vulnerability at the Valparaíso interface.

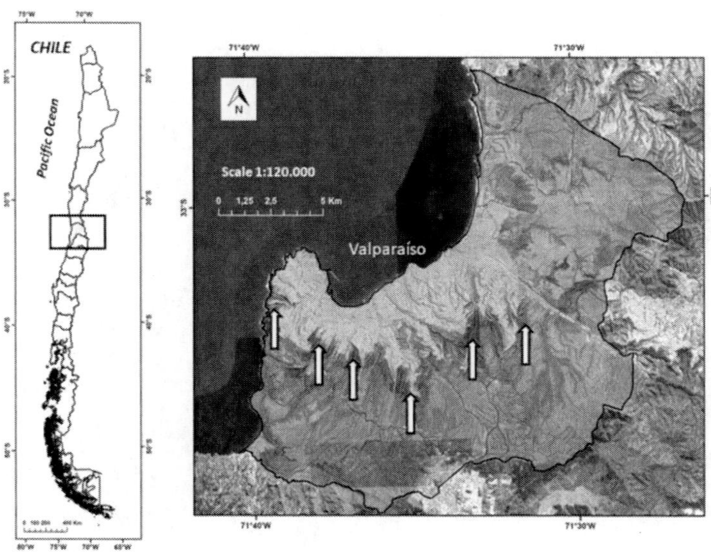

Figure 6. Result obtained from the vulnerability analysis for the urban forest interface of Valparaíso. The arrows indicate the highest values in the count of the variables associated with risk, danger and potential damage.

In relation to the methodology described here, the process of determining critical zones of vulnerability based on this criteria scheme, can be completely applicable to other latitudes or regions that present forest fire problems at the interface, considering the adaptations necessary to adequately represent the specific aspects in other Mediterranean regions. The social behavior of the population at risk is an important aspect to evaluate and consider for the design of strategies in pre-attack plans. This investigation also was carried out through interviews with representatives of the neighborhood councils located in the areas with the least availability of infrastructure and the greatest difficulty in land access.

This information and the general criteria were applied to a previous zoning of the areas most exposed to the impact of forest fires, and which surround Valparaíso in its consolidated urban area. In these areas the presence of dense vegetation predominates, corresponding to thickets and native forest (sclerophyllous forest), together with the presence of exotic tree species, such as pine (*Pinus radiata* D. Don), aromo (*Acacia dealbata* Link) and eucalyptus (*Eucalyptus* spp.). These last three are promoters of the propagation speed in the presence of strong winds, thus becoming a factor of high danger for the advance of fire in inhabited areas of the interface in Valparaíso.

Figure 7. Geographical representation of 27 critical areas in the urban forest interface for the construction of pre-attack plans.

For the determination of the coverage areas for the preparation of pre-attack plans, it was necessary to divide the interface territory into 27 zones according to the unique attributes of each: density and proximity of houses, proximity to forested areas, form of hill slopes and local attributes of climate and local connectivity. These criteria are also useful and necessary to establish a pre-zoning in other interface areas. In Figure 7 two relevant data are extracted: the identifier of the area to be treated for the pre-attack plans and the surface of each one, with the aim of dimensioning the scope and technical feasibility of each preparatory action.

In relation to these 27 classified areas of special interest, a second, more specific analysis was carried out, now related to three criteria that are later combined with the vulnerability analysis: they are criteria associated with the response of the fire in the advance of combustible material (speed of propagation and caloric intensity), population vulnerability and difficulty in combat. These aspects form, together with the territorial analysis of protection priorities, the basis for proposing local pre-attack plans for each of the Valparaíso areas, in the context of house defending at urban interface. The following table shows the additional indicators that were collected according to field experience, from the interview with representatives of the communities living in the different areas of these interface, and from the information gathered on topography, vegetation characteristics and the local climate, with a similar approach to vulnerability analysis but now with an emphasis on the operational aspect for the organization of combat devices and territorial planning in the preparation of preventive and pre-attack plans.

An essential step in preparing the diagnosis of the territorial units under threat of fire (in this case, there are 27 units), is the characterization of the dwellings, from the point of view of the criteria described in Table 2. This step is strictly necessary for the elaboration of a detailed cadastre, because the omission results in possible errors in the allocation of the workload and the classification of the risk against emergencies. For this purpose, a detailed field trip was carried out with the support of high-resolution aerial photos and Google Earth images to identify and separate groups of houses. This activity allowed the identification of 4 groups or types of houses that form the urban forest interface in Valparaíso. For example, light material homes located in areas of high vulnerability, require shorter arrival times of the first resource, which can be land or air. Normally in this type of landscape where there is a predominance of groups of houses at the interface and with the presence of very irregular topography, the first attack is carried out by plane or helicopter. In this sense, the references of Castillo (2015) were taken on a study in response times to different types of forest fires and their levels of conflict. In pre-attack plans this information is extremely important for decision making.

Table 2. Definition of specific criteria within each of the zones defined for the pre-attack plans. Example applied to Valparaíso, possible to apply in other areas and regions with similar interface fire problems

General criteria	Specific criteria	Factor	Score
Hazard (250 points)	Fire spread	Very low	12.5
		Low	37.5
		Medium	62.5
		High	100.0
		Very high	125.0
	Fire intensity	Very low	12.5
		Low	37.5
		Medium	62.5
		High	100.0
		Very high	125.0
Population vulnerability (400 points)	Materials	Solid	30.0
		Mixed	70.0
		Light material	100.0
	Density	An isolated house	20.0
		2 join houses	30.0
		A house group	50.0
	Distance of main roads	0-500 m.	10.0
		500-1,000 m.	20.0
		1,000-2,000 m.	30.0
		> 2,000 m.	50.0
	Distance of secondary roads	0-500 m.	50.0
		500-1,000 m.	30.0
		1,000-2,000 m.	20.0
		> 2,000 m.	10.0
	Potential affected perimeter around houses	0-500 m.	30.0
		500-1,000 m.	50.0
		1,000-2,000 m.	80.0
		2,000-3,000 m.	95.0
		3,000 y más m.	100.0
	Water supply (fixed point)	> 3 taps	10.0

General criteria	Specific criteria	Factor	Score
		2 taps	20.0
		1 tap	30.0
		Without taps	50.0
Fire fighting difficulty (350 points)	Quality of accesses	Concrete pavement	40.0
		Gravel road	60.0
		Dirt road	100.0
	Amount of accesses	> 3	10.0
		2	20.0
		Unique access	70.0
		Without access	100.0
	supply points (water, pools, others)	> 3	10.0
		2	20.0
		1	70.0
		Unavailable	100.0
	facility for coordination station installation (command)	Good	10.0
		Acceptable	20.0
		Unavailable	50.0

Due to the high variability observed in the housing sectors in the interface area, the first necessary step for the assignment of standardized scores was the realization of a cadastre with the support of satellite images and field evaluations, the result of which showed four different situations:

Situation 1: Consolidated housings sectors of a high socio-economic level, characterized by the following indicators: total area in the study area: 99 ha; housing density (number of houses per hectare): 26; average surface of houses: 120 m^2; Average number of inhabitants per house: 2 adults and two children. Construction materials (in order of priority): brick-concrete-clay tiles-wood-cement. Surrounding vegetation: gardens with irrigation and ornamental trees.

Situation 2: Partially consolidated housing sectors of a medium to low socioeconomic level. Total area in the study area: 414 ha; housing density (number of houses per hectare): 11; average surface of houses: 48 m^2; average number of inhabitants per house: 3 adults and 5 children; construction materials (in order of priority): wood-volcanite; surrounding vegetation: thickets and grasslands.

Situation 3: Unconsolidated housing sectors of a low socioeconomic level, with the following antecedents: total area in the study area: 220 ha;

housing density (number of houses per hectare): 72; average surface of houses: 52 m^2; average number of inhabitants per house: 3 adults and 4 children; construction materials (in order of priority): slate-volcanite-wood-brick; surrounding vegetation: thickets and grasslands.

Situation 4: Consolidated housing sectors of a medium socioeconomic level, preferably rural, with the following antecedents: total area in the study area: 4,656 ha; housing density (number of houses per hectare): 1; average surface of houses: 68 m^2; average number of inhabitants per house: 3 adults and 4 children; construction materials (in order of priority): cement-wood-brick-concrete; Surrounding vegetation: thickets, grasslands and agricultural crops (fruit trees, cereals and vegetables).

The total extension of the interface, including the four situations indicated above, is 5,390 ha, which corresponds to 24.26% of the surface of the study area. Photographs are shown in Figure 8, showing common examples of the four interface situations just described.

Figure 8. Urban-forest interface situations present in the study area (Valparaíso, Chile).

With all the previous background on defining general and specific criteria for areas of vulnerability, and also the variables associated with the characterization of specific areas for pre-attack plans in interface fires, a map of priorities for attention was generated accordingly. of emergencies, through a process of adding layers in the Geographic Information System and a subsequent classification into 3 classes, to separate the effect of each of the variables within the total set. The result obtained was a priority map for the pre-attack plans, which contains a classification in priorities: 1 (first), 2 (second) and 3 (third), for planning fire fighting equipment and resources, in circumstances of multi-occurrence and budget constraint in the first response (initial attack time). For this purpose, pre-attack plans must be designed through an accurate inventory of available resources, not only in aircraft and land resources, but also in parking lots, airports for the departure and arrival of planes and helicopters, sources of gasoline and water supply.

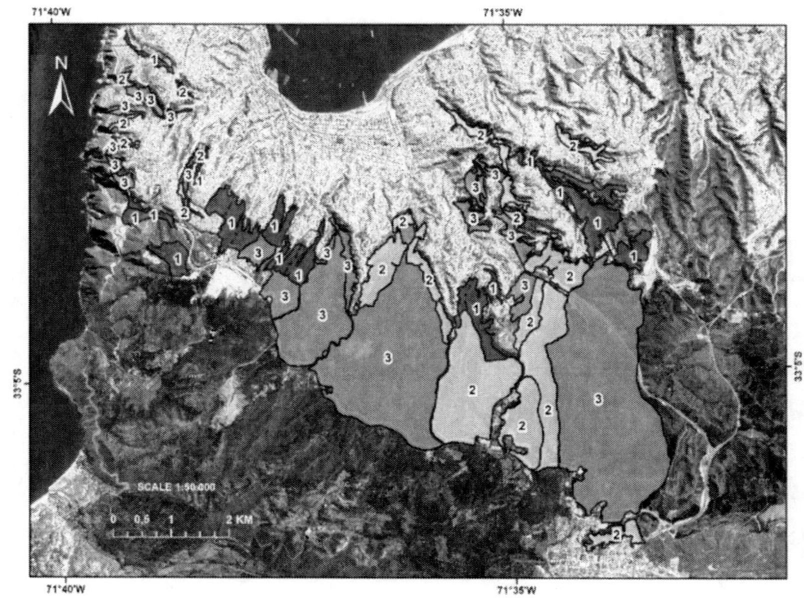

Figure 9. Priority map for emergency responses in pre-attack plans. First response to a potential fire at the wildland urban interface. Diagnosis for the 27 areas identified in Valparaíso.

In consideration of these results, a map is therefore obtained to determine priority areas for the allocation of resources to combat wildfires in areas of urban forest interface, and which considers not only environmental and vegetation variables, but also other variables. Operative aspects such as the difficulty of access, the potential intensity of the fire in the areas of greatest danger and also the infrastructure, should be considered for the evaluation of potential zones for the first attack. In this particular case, the Valparaíso Commune presents many areas with very high slopes. In each case study, and also in other Mediterranean regions where it would be possible to apply this methodology, it is extremely important to consider the opinion and participation of the communities that are directly exposed to the impact of a forest fire near their homes. The Valparaíso experience represents a good example to demonstrate that it is possible to build a pre-strategy for the start of the forest fire season, using practical indicators that allow for a more precise evaluation of the interface areas, which normally only consider the place of fire start and a prevention criterion not necessarily associated with the previous level of vulnerability.

Conclusion

Wildfires in the urban-forest interface areas of the Valparaíso commune in the last five years have caused significant damage, affecting numerous homes and causing the loss of human lives, a situation that is expected to repeat in the future, as a result of the climate change and increased vulnerability of peri-urban sectors, which will significantly affect the presence of fires with extreme behavior, difficult to control and greater potential damage, leading to events of catastrophic characteristics.

Consideration of the greatest number of variables and specific criteria associated with the detailed description of small areas, such as groups of houses, their structural characteristics and location, are extremely relevant data for the planning of priority areas and consequently the allocation of financial resources for prevention and participation campaigns with communities. In this particular example, Valparaíso has a great tourist activity. However, in this study it was chosen to concentrate the research on those areas that do not necessarily have high tourist activity, since they are remote areas in the mountains with forests and irregular areas with difficult access for tourist infrastructure.

The determination of Protection Priorities was based on the integration of the results from three specific analyses: Risk, Danger, and Potential Damage or Vulnerability. These results are integrated with a second, more detailed analysis of on-site zoning that resulted in 27 interface units or zones. In them, a second multi-criteria analysis was applied based, among other aspects, on the intensity of the fire, aspects of accessibility and difficulty for combat, which allowed, consequently, to be able to classify and territorially qualify those sectors that are most susceptible to the spread of fire, to the damages and impacts that fires can generate and the degree of difficulty in their control and extinction, through preparatory plans for emergency care, called pre-attack plans.

The antecedents obtained by the determination of protection priorities from the analysis of the danger, population vulnerability and difficulty for combat, allow us to characterize very well the problems that forest fires cause, regarding the expected behavior of the fire, the losses and impacts that these can cause, generate, and form an important basis for the elaboration of strategic protection plans in order to prevent, combat and mitigate the effects of fire in the study areas.

ACKNOWLEDGMENTS

The author of this chapter wishes to express his sincere thanks to the National Forest Corporation of Valparaíso (Chile) for the contribution in geographic data, surveys and cartographic material that was useful for the development of the investigation. Special gratitude to Juan Atienza and Luis Correa. Also to Project A/017536/08 executed in 2010 in Valparaíso, whose results allowed to know in more detail the characteristics of the urban interface.

References

[1] Rodríguez y Silva, F. (2000). Defense Plans in Spain - Example from Andalusia (In Spanish). In: *Defense against Forest Fires - Foundations and Experiences*. McGraw Hill. Madrid.

[2] Rodríguez y Silva, F. (2009). Defense Plans in Spain - Procedures and Methodology for Preparation. In: *Defending Against Wildfires - Fundamentals and Experiences* (In Spanish). Second Edition. McGraw Hill. Madrid. 289-322.

[3] Julio, G. (2014). *Fire Management Fundamentals. VI Edition* (In Spanish*)*. University of Chile. School of Forest Science. Santiago.

[4] Rodríguez y Silva, F. (2004). Development of a Model for the Economic Evaluation of Forest Fire Defense Plans in Mediterranean Forest Ecosystems (In Spanish). In: *Proceedings II International Symposium on Policy, Planning and Economics of Forest Fire Protection Programs.* Córdoba, Spain.

[5] Rodríguez y Silva, F., Molina, J., Herrera, A., and Zamora, R. (2008). The Impacts of Fires and the Socioeconomic Vulnerability of Forest Ecosystems. A methodological approach using Remote Sensing and Geographic Information Systems (In Spanish). In: *Proceedings III International Symposium on Policy, Planning and Economics of Forest Fires*. Carolina, Puerto Rico.

[6] Castillo Soto, M. E. (2013). *Integration of variables and territorial criteria to support protection against forest fires. Pilot area: Valparaíso-Central Chile* (In Spanish). University of Chile.

[7] Castillo, M. E., & Rodriguez, F. (2015). Determining response times for the deployment of terrestrial resources for fighting forest fires. A case study: Mediterranean–Chile. *International Journal of Agriculture and Natural Resources*, 42(1), 97-107.

[8] Rehm, R. G., & Mell, W. R. (2009). A simple model for wind effects of burning structures and topography on wildland–urban interface surface-fire propagation. *International Journal of Wildland Fire*, 18(3), 290-301.

[9] Porterie, B., Consalvi, J. L., Loraud, J. C., Giroud, F., & Picard, C. (2007). Dynamics of wildland fires and their impact on structures. *Combustion and Flame*, 149(3), 314-328.

[10] Pyne, S., Andrews, P., and Laven, R. (1996). *Introduction to wildland fire.* 2nd ed. New York, NY: John Wiley & Sons. 808 p.

[11] Beebe, G. S., & Omi, P. N. (1993). Wildland burning: the perception of risk. *Journal of Forestry*, 91(9), 19-24.

[12] Julio, G. (1992). *Protection Priority Determination Method* (In Spanish). University of Chile. School of Forest Science. General Report, 10-28.

[13] Grayzeck-Souter, S. A., Nelson, K. C., Brummel, R. F., Jakes, P., & Williams, D. R. (2009). Interpreting federal policy at the local level: the wildland–urban interface concept in wildfire protection planning in the eastern United States. *International journal of wildland fire*, 18(3), 278-289.

[14] Kundell, J. E., Myszewski, M., & DeMeo, T. A. (2002). Land use planning and policy issues. *The Southern Wildland-Urban Interface Assessment*, 1992(97), 53.

[15] Hammer, R. B., Radeloff, V. C., Fried, J. S., & Stewart, S. I. (2007). Wildland–urban interface housing growth during the 1990s in California, Oregon, and Washington. *International Journal of Wildland Fire*, 16(3), 255-265.

[16] Davis, J. B. (1990). The wildland-urban interface: paradise or battleground? *Journal of forestry*, 88(1), 26-31.

[17] Radeloff, V. C., Hammer, R. B., Voss, P. R., Hagen, A. E., Field, D. R., & Mladenoff, D. J. (2001). Human demographic trends and landscape level forest management in the northwest Wisconsin Pine Barrens. *Forest Science*, 47(2), 229-241.

[18] Massada, A. B., Radeloff, V. C., & Stewart, S. I. (2011). Allocating fuel breaks to optimally protect structures in the wildland–urban interface. *International Journal of Wildland Fire*, 20(1), 59-68.

[19] Rodríguez y Silva, F., Julio, G., Castillo, M., Molina, J., Herrera, M., Toral, M., Cerda, C. and González, L. (2010). *Application and adaptation of the SEVEIF Model for the socioeconomic evaluation of*

the impact of forest fires in the Province of Valparaíso, Chile (In Spanish). Agencia Española de Cooperación Internacional para el Desarrollo (AECID). 52p. ISBN 978-84-693-0740-3.

[20] Bowman, D. M., Moreira-Muñoz, A., Kolden, C. A., Chávez, R. O., Muñoz, A. A., Salinas, F., ... & Borchers, N. (2019). Human–environmental drivers and impacts of the globally extreme 2017 Chilean fires. *Ambio*, 48(4), 350-362.

BIOGRAPHICAL SKETCH

Miguel E. Castillo Soto

Affiliation: University of Chile

Education: Forestry Engineer, Graduate in Mathematical Cartography, MSc. Geography, PhD. Forestry

Business Address: Av. Santa Rosa 11.315. La Pintana. Santiago of Chile

Research and Professional Experience:

Professor and Researcher of the Forest Fire Laboratory (Faculty of Forest Sciences and Nature Conservation, University of Chile, from 1995 to date). Author of national (SciELO) and international (Wos-ISI) publications related to various areas of fire management: fire ecology and restoration, civil protection, urban-forest interface, as well as the design and implementation of fire management tools for taking decisions, and the mathematical and cartographic modeling of phenomena related to the behavior of fire, its economic, social and territorial consequences. Experience since 1995 in Mathematical Cartography and geomatic applications. Responsible for various research projects based on competitive funds (FONDEF, CONAF, others). He has been professor of this Faculty in the line of Fire Management, Geographical Information Systems (GIS), and lately in the Graduate School teaching subjects related to Fire Ecology and Geomatics applied to forest fire control. Professor of the Doctorate of Silvo-agricultural and Veterinary Sciences of the University of Chile and the Master in Wild Areas and Nature Conservation. Professor of the Master in Environmental Management and Planning. Member of the Asian Council of Science Editors, and arbitrator of various international magazines related to modeling in forest fires. He has obtained specializations in Remote Sensing and GIS in Spain, Portugal and Colombia. Consultant of national and international projects and rapporteur of various training courses with special emphasis on the application of cartographic tools for the integrated study of the territory. Expert in forest expertise and investigation of causes and analysis of large-scale claims and historical behavior of fire. Presence in national and international media. As a consultant, he has investigated the main forest fires that occurred in Chile, especially the 2017 mega-fires in Chile. He is currently the Director of the Department of Forest Management and Full-time Professor of Undergraduate and Postgraduate subjects. Researcher attached to the Forest Fire Laboratory of the University of Chile.

Professional Appointments: Director (since 2019) of Department of Forest Management and Environment. Faculty of Forest Science and Nature Conservation. Associate Professor. Researcher at the Forest Fire Laboratory of the University of Chile since 1995.

Publications from the Last 3 Years:

1) Castillo, M. (2019). Riesgos de incendio forestal. En: *Catástrofes Mistas: uma perspectiva ambiental*. Serie Riscos e catástrofes. [Forest fire risks. In: *Mixed Catastrophes: an environmental perspective*. Risk and disasters series.] ISBN 978-989-26-1902-6. doi https://doi.org/10.14195/978-979-26-1901-9. 520p.
2) Quintanilla, V.; Castillo, M.; Garfias, R. (2018) Análisis espacio-temporal a causa de incendios forestales del paisaje de bosque esclerófilo de Chile Mediterráneo. En Libro: *Bosque Mediterráneo y Humedales: Paisaje, Evolución y Conservación* [Spatio-temporal

analysis due to forest fires of the sclerophyllous forest landscape of Mediterranean Chile. In Book: *Mediterranean Forest and Wetlands: Landscape, Evolution and Conservation*]. pp.300-304. Almud Ediciones. Castilla la Mancha. España.

3) Castillo, M.; Garay, R.; García, P.; Radeloff, V.; Espinoza, P. (2017) The Risk of wildfires in Chile. In: *Wildfires: Perspectives, Issues and challenges of the 21st century*. Chapter 3, 55-80.
4) Castillo, M.; Garay, R.; Tapia, R.; Garfias, R.; Orell, M. (2020). Metodología de evaluación de infraestructuras críticas en zonas de riesgo de incendios forestales. [Critical infrastructure assessment methodology in forest fire risk areas]. *Academic Repository. University of Chile.* URI: http://repositorio.uchile.cl/handle/2250/173421 ISBN: 9789561911628. Santiago de Chile 249p.
5) Garfias, R.; Castillo, M.; Vita, A.; Bown, H.; Zúñiga, P.; Ruiz, F.; Caldentey, J. (2019). *Funciones de biomasa y estimación de carbono fijado por las principales especies del bosque esclerófilo de las regiones de O'Higgins y del Maule*. Libro. Fondo de Investigación del Bosque Nativo. [*Biomass functions and estimation of carbon fixed by the main species of the sclerophyllous forest of the O'Higgins and Maule regions*. Book. Native Forest Research Fund.] ISBN 978-956-19-1153-6. 62p.
6) Garcia-Chevesich, P.; Martínez, E.; García, A.; Castillo, M.; Garfias, R.; Neary, D.; Pizarro, R.; Valdés, R.; González, L.; Venegas-Quiñones, l.; Magni, C. (2019). Formation of Post-Fire Water Repellent Layers on Nothofagus glauca (Hualo) Forests, After the Historical "Las Máquinas" Wildfire in South-Central Chile. *American Journal of Environmental Sciences*. doi: 10.3844/ajessp.2019.
7) Garfias, R.; Castillo, M.; Allende, C.; Ruiz, F. (2019). Un método de zonificación ambiental para el establecimiento de especies con uso pastoril en Chile Central. [An environmental zoning method for the establishment of species with pastoral use in Central Chile]. *Interciencia* 44(1).

8) Plaza, A.; Castillo, M. Germination rates of four Chilean forest trees seeds: Quillaja saponaria, Prosopis chilensis, Vachellia caven, and Caesalpinia spinosa. *F1000 Research* 7:1446.. https://doi.org/10.12688/f1000research.16091.1.
9) Garfias, R.; Castillo, M.; Ruiz, F.; Vita, A.; Bown, H.; Navarro, R. Remanentes del bosque esclerófilo en la zona mediterránea de Chile central: caracterización y distribución de fragmentos. [Sclerophyllous forest remnants in the Mediterranean area of central Chile: characterization and distribution of fragments]. *Interciencia* 43(9):655-663.
10) Molina, J-R.; Moreno, R.; Castillo, M.; Rodríguez y Silva, F. (2017). Economic susceptibility of fire-prone landscapes in natural protected areas of the southern Andean Range. *Science of the total Environment* V. 619–620:1557-1565.
11) González, C.; Castillo, M.; García-Chevesich, P.; Barrios, J.(2017). Dempster-Schafer theory of evidence: A new approach to spatially model wildfire risk potential in central Chile. *Science of the Total Environment* 613-614 (2018) 1024-1030.
12) Castillo, M.; Saavedra, J.; Brull, J. (2019). Fire severity in mega wildfires in Chile 2017. Actions to improve de protection system. *Territorium* 26(1), 5-18. doi: https://doi.org/10.14195/1647-7723_26-1_1.
13) Machuca, J.; López, M.; Castillo, M. (2018). Diseño de un modelo de cuantificación de calidad hídrica en humedales. [Design of a model for the quantification of water quality in wetlands]. *Investigaciones Geográficas* (22):23-44.
14) Garay, R.; Tapia, R.; Castillo, M.; Fernández, O.; Vergara, J. (2018). Habitabilidad de edificaciones y ranking de discriminación basado en seguridad y sustentabilidad frente a eventuales desastres. Estudio de caso: Viviendas de Madera. [Habitability of buildings and discrimination ranking based on safety and sustainability against eventual disasters. Case Study: Wooden Houses]. *REDER* (2,2), 28-45. ISSN 0719-8477.

In: Critical Issues and Analysis … ISBN: 978-1-53618-738-0
Editors: A. Cablé and B. Manescau

Chapter 7

A Fire Station Solar Energy Retrofit Model for Environmental Community Risk Reduction

Scott Somers*

Watts College of Public Service and Community Solutions,
Arizona State University, Phoenix, Arizona, US

Abstract

This chapter discusses how fluctuations in energy costs strain the financial security of cities, and highlights how environmental degradation due to energy consumption affects community health and safety, which increases demand for local public safety services. Cities are looking to cut emissions by installing renewable energy systems in new buildings and retrofitting existing structures. With responsibility for all-hazards community risk reduction, the fire service should be a leader in comprehensive sustainability planning. A case study of an evaluation of local fire stations is presented as a starting point to explore how solar energy retrofit projects are being identified to save money, reduce air

* Corresponding Author's Email: ssomers@asu.edu.

pollution, improve public health and safety, and boost the city's reputation for leadership in environmental sustainability. This model can be replicated by any fire service organization to assess the potential for solar energy retrofits.

Keywords: solar, renewable energy, community risk reduction

INTRODUCTION

There are complex linkages of technology and infrastructure that affect climate change [1]. Climate change represents a critically important issue for the safety, health, and welfare of humankind with a wide array of potentially disastrous effects threatening cities, while buildings have a significant impact on energy use and the environment [2, 3]. Municipalities construct fire stations within population centers to meet growth and service demand. Fire stations are unique compared to most municipal facilitates since they systematically operate 24 hours per day and 365 days per year. This case study evaluates City of Mesa, Arizona fire stations as a local starting point to explore how solar energy retrofit projects are being identified to save money, reduce air pollution, improve public health and safety, and boost the city's reputation for leadership in environmental sustainability.

According to statistics compiled by the U.S. Environmental Protection Agency, buildings account for 72% of the nation's electricity consumption, 39% of carbon dioxide emissions, and one-third of green house gas emissions [4]. Ensuring that buildings meet strict environmental standards presents one of the most accessible opportunities to make deep cuts in emissions. Many cities are embracing Leadership in Energy and Environmental Design (LEED) green building certification for new municipal facilities, with some buildings constructed to utilize renewable energy sources such as solar photovoltaic panels. One challenge concerns how to identify opportunities to retrofit existing building stock, much of which is energy inefficient, yet will still be in use for many decades to come.

Located in an urban area of over 4 million people, Mesa is susceptible to heat island effect, temperature inversions, particulate pollution, dust storms, high ozone, wildfires, severe drought, and other hazards. Arizona temperatures increased 0.273-degrees Fahrenheit over the past decade, the fourth highest increase of any state in the U.S. [5]. The city owns more than 200 buildings with over 3 million square feet of space. Utility bills for fiscal year 2009-2010 were $13 million for 140 million kilowatt hours of energy [6].

Specifically, this article documents efforts of the Mesa Fire Department to identify existing department facilities suitable for solar energy retrofits. It describes the institutional motivations and political support for environmental sustainability, which is essential to such efforts, and discusses both the financial hurdles and incentives to solar energy adoption in Arizona. The study was originally conducted in 2011-12 to find ways to reduce departmental operating costs, but also to achieve the environmental goals set by city council. Finally, the method used to inventory particulate and green-house gas emissions from fire department facilities and identifying solar ready roof area suitable for a solar photovoltaic retrofit is presented.

INSTITUTIONAL MOTIVATIONS

Air Quality and Public Health

Poor air quality is a pressing environmental issue, especially for densely urbanized regions. Once renowned for its clean air, Arizona was considered a destination for people suffering from chronic lung diseases or environmental allergies. As the region grows, however, Phoenix struggles to maintain acceptable air quality standards. Metropolitan Phoenix is located in a valley surrounded by mountains, which predisposes the area to frequent temperature inversions. Particularly during the winter months, "an inversion is an atmospheric condition caused by increasing temperature with elevation, resulting in a layer of warm air preventing the rise of cooler air trapped beneath" [7]. During the day, as the desert floor heats up, trapped polluted air spreads and particulate matter rises creating a visible haze [8].

The Phoenix-Mesa-Scottsdale region appears on lists for most polluted cities in the American Lung Association State of the Air report [9]. State of the Air is a national air quality "report card" that assigns letter grades to counties across the country and ranks cities and counties on three measures of the most widespread air pollutants – ozone and particle pollution. Ozone is a colorless, highly reactive gas that is the principle component of smog [10]. Particle pollution, also known as particulate matter or PM, "is a complex mixture of extremely small particles and liquid droplets. Particle pollution is made up of a number of components, including acids (such as nitrates and sulfates), organic chemicals, metals, soil or dust particles" [11]. The region ranked 19th for ozone1, 2nd for year-round particle pollution, and 24th for short-term particle pollution.

As a whole, Maricopa County does better with particle pollution receiving a "B" grade in the State of the Air report. However, the Phoenix-Mesa-Scottsdale area, where the bulk of the Maricopa County population is concentrated, does not fare as well. In a January 25, 2011 press release, the Arizona Department of Environmental Quality announced the state had withdrawn its particulate matter air quality plan for Maricopa County to avoid a finding on non-compliance by the Environmental Protection Agency. The plan calls for significant reductions in particle matter of 10 microns or less, commonly known as PM10 [12].

Epidemiological studies conducted in Arizona established a strong correlation between air quality and public health, particularly among at-risk populations. Mar, et al. [11] evaluated the association between mortality outcomes in the elderly and particulate matter. Cardiovascular mortality was found to be significantly associated with carbon monoxide (CO), nitrogen dioxide (NO2), sulfur dioxide (SO_2), particulate matter 2.5, particulate matter10, and ozone levels in atmosphere. A study by Arizona Department of Environmental Quality in collaboration with the Arizona Department of Health Services and Arizona State University's Center for Health Information Research examined the connection between particulate matter10 and incidence of asthma among children. It found that the asthma events among children aged 5-18 years increased 13.7% when the daily average of particulate matter10 rose to the 75th percentile. Furthermore, neighborhoods with lower socioeconomic status, higher populations of Latino immigrants, and higher proportions of renters have been found to have greater exposure to air pollutants [14].

Climate Change and Fire

Global climate change has the potential to exacerbate pollution-related problems in Arizona and has been recognized by scientists as a serious public health and environmental concern. As atmospheric concentrations of greenhouse gasses rise, global temperatures are increasing. The Intergovernmental Panel on Climate Change predicted that climate change may lead to more extremes of heat and cold; increase the prevalence of climate-sensitive diseases including malaria, dengue fever, yellow fever, and encephalitis; and diminish air quality directly as hotter days facilitate ozone formation and indirectly affect the concentration of PM in the air by affecting natural sources of PM such as wildfires and dust from dry soils [15]. An analysis of the effects of global climate change on human health and welfare by the U.S. Climate Change Science Program concluded that the anticipated impact of global warming on the intermountain west will include early snowmelt, degraded air quality, more intense urban heat island effect, wildfires, heat waves, and drought [1].

The Arizona Climate Change Advisory Group [16] observed that over the past 50 years the climate in the western United States warmed, on average, 1.4 degrees Fahrenheit. It cited the intergovernmental climate models that predict annual mean temperatures in the American southwest could increase by as much as 14 degrees Fahrenheit before the end of the century. The National Weather Service Forecast Office in Phoenix reported that high and low temperatures have been trending warmer since records began in 1896. Arizona greenhouse emissions are rising faster than the national average. A climate change group survey of Arizona emissions documented an increase of 51% from 1990 to 2000, compared to only 23%

nationally during this period [17]. The effects of global warming on Arizona are expected to be severe and include an impact on the state's water supply as high temperatures increase evaporation off reservoirs; warmer temperatures may intensify the urban heat island effect, especially in the metropolitan Phoenix and Tucson areas, worsening Arizona's air pollution problems; stronger summer storms and earlier mountain snowmelt could contribute to greater flash flooding; and hotter, drier temperatures may produce larger, more intense wildfires and a longer wildfire season [18].

A study by Westerling, Hidalgo, Cayan, & Swetnam [19] observed since 1987, the wildfire season in the western states became 2.5 months longer due to climate changes brought about by global warming. The paper analyzed 1,166 large forest fires in excess of 1,000 acres from 1970 to 2004 and noted that, in the second half of that period, there were four times as many wildfires, and 6.5 times as much acreage burned. A search of the Disaster Information database on the FEMA website logged 70 disaster declarations in Arizona since 1966. Forty-seven wildfire events are recorded, all occurring after 1987. In fact, after a decade of drought and warmer temperatures, the Arizona wildfire season had its earliest start on record in February 2006, and the state's two largest wildfires, the 2002 Rodeo-Chediski Fire and the 2005 Cave Creek Complex Fire, collectively consumed over 750,000 acres [16, 18]. In June 2011, the Wallow Fire surpassed Rodeo-Chediski as the largest wildfire in Arizona history burning 538,049 acres and destroying 72 buildings. Wildfires are themselves producers of emissions [20]. Smoke produced by wildfires contains gasses and PM that contribute to local and regional haze in Arizona [21].

Arizona Greenhouse Gas Emissions

The Arizona Climate Change Advisory Group [16] observed that over the past 50 years the climate Data from the Final Arizona Greenhouse Gas Inventory and Reference Case Projections, 1990-2020 shows that electricity use accounts for nearly 40% of the state's gross emissions, "slightly higher than the national share of emissions from electricity production" [17]. Coal is the largest source of electricity in Arizona, accounting for 35.5% of production, or 40 million megawatt hours (MW•h), followed by natural gas (31.1%, 36 MW•h), and nuclear (27.4%, 30 MW•h); oil accounts for 0.1% of Arizona energy production [22, 23]. Emissions from coal-fired power plants contain 84 of 187 hazardous air pollutants identified by the EPA as posing a threat to human health and the environment and are a major source of sulfur dioxide, oxides from nitrogen, and PM [24].

While the Arizona Climate Change Advisory Group raises concerns over greenhouse gasses emissions, there is a positive trend worth noting: During the 1990s, Arizona electricity demand grew at a rate of 4.0% per year, while electricity emissions grew 3.3% annually, reflecting a decline in emissions per kW•h. This decline was due largely to the rapid growth of new natural gas generation, and to a lesser extent increases in nuclear generation [17]. Since natural gas and nuclear energy are "cleaner" than other forms of fossil fuels, this suggests that an increased use of renewable energy sources may not only retard the growth of emissions per kilowatt-hour, but also decrease gross emissions in the state.

Teruyuki Kobayaski, former Fire Chief of the Tokyo, Japan Fire Department, feels that the fire service can make a contribution to reducing the effects of greenhouse gases and improving its carbon footprint by seeking efficiency by taking steps to conserve energy and monitoring departmental environmental protection initiatives, and reducing fuel consumption [25]. Indeed, local government agencies can make meaningful contributions to reducing greenhouse gasses as they "have considerable authority over land-use planning, and waste management and can play an important role in . . . energy consumption" [26].

Political Leadership

The City of Mesa has continuously elevated its leadership role in stewardship of the environment by leading and participating in regional and local initiatives. The city is an organizing member of the Sustainable Cities Network, a regional initiative of Arizona State University, city, county and tribal leaders for sharing knowledge and coordinating efforts to understand and solve problems related to front-line challenges of municipal sustainability. The city's Deputy Director of Environmental and Sustainability serves as chair of the Sustainable Cities Network solar committee.

In 2009, then-Mayor Scott Smith became the 1,000th signatory to the U.S. Conference of Mayors Climate Protection Agreement. The agreement aligns with Kyoto Protocol standards, which call for a reduction in carbon emissions by 7% below 1990 levels [27]. Five Mesa City Council Strategic Initiatives provide a framework for the City strategic plan; these initiatives call for regional leadership that partners to find common solutions to regional challenges that are innovative and sustainable, and seek to address quality of life issues in a manner that is sustainable and environmentally friendly. Toward these goals, the city has participated in several local sustainability projects including a public-private partnership to install concentrated solar photovoltaic technologies at the city's Central Arizona Project Water Treatment Plant; a partnership with Arizona State University to integrate research of solar and other alternative energy resources with Mesa's electric distribution utility; the Transportation Department has installed photovoltaic panels to provide lighting for bus shelters to obviate the need for electrical service; and the fire department sought LEED certification for fire stations 219 and 220.

The fire department's strategic plan defined the organization's desired future state, its purpose, and declares the values that should guide decision-making. To achieve that mission for the future, ensure alignment with the city's mission, and provide guidance for operational components of the plan, the department focuses on seven key areas; among these is attention to neighborhood and environmental vitality. Part of the vision for this key area is to promote environmental awareness [28]. The fire department's strategic plan purposefully aligns with the city's voter approved general plan. Air quality management and central theme in the general plan. Key policies in the general plan focus on improving air quality and conservation of energy resources by encouraging solar conscious design and apply energy conservation techniques in the development and operation of municipal facilities [29]. The City's 2040 General Plan notes the expansion of the City's solar technology portfolio. Between 2010 and 2013 solar energy generation increased to 841 kW (the equivalent of 85 households for a year) providing energy cost savings for the City operations [30].

Financial Incentives

A key factor in the decision pursue this project was the city's continuing budgetary constraints, which are exacerbated, in part, by fluctuating energy costs. The years during the Great Recession were difficult economically for the Mesa, as they had been for many American cities. The decline in the national economy was felt directly by the city with a dramatic decrease in local sales tax revenues and was shared throughout the State through declines in state shared revenue receipts. The extent of the budget reductions necessitated by revenue declines was significant. Ultimately, the approved fiscal year 08/09 budget was reduced by $29 million and $60 million of anticipated expenditures were eliminated from the 09/10 budget [31]. As the 2009-2010 fiscal year progressed, a variety of continuing economic impacts were recognized. General fund projections were forecasted to decrease another $16 million, with additional reductions of over $3 million anticipated for years 10/11 [6]. City management continues to explore better, more efficient ways of providing services. This includes identify energy conservation measures that help reduce energy consumption, decrease operating expenses, and ease the maintenance requirements for city buildings.

The high net installed cost of solar energy in relation to grid-supplied power remains a major impediment to its wider adoption [32]. Economic studies of solar power continue to conclude that, although costs are declining, without significant federal, state, and utility incentives, the installed costs of solar far exceeds the net present benefit [32, 33]. As non-taxable entities, local governments cannot capture the full range of tax incentives available to private corporations, assuming they purchase the system. The use of 0% federal bond programs such as the Clean Renewable

Energy Bond and the qualified energy conservation bond as well as the sale of solar renewable energy credits on the utility market and reduced electricity costs, offer some financial benefits. However, there is uncertainty about the sustainability of the Clean Renewable Energy Bond program, and the value of solar renewable energy credits is dependent upon market conditions; specifically, whether a state has a mandatory or voluntary renewable energy standard [34]. Ultimately, these incentives are necessary, but not sufficient, to offset the upfront cost of installing photovoltaic panels.

Third-party ownership offers an alternative to self-financing that can maximize the use of incentives by introducing a private, taxable entity into a solar power partnership. A third-party ownership arrangement through a power purchase agreement offers many benefits including providing predictable energy costs, captures federal incentives like the Solar Investment Tax Credit and Modified Accelerated Cost Recovery System, defrays up-front capital costs, and transfers responsibility for operation and maintenance to the solar service provider. Solar leases offer similar benefits however, since the public agency would be leasing the solar equipment from the solar service provider rather than purchasing the electricity, the agency takes on responsibility for system performance. The risk is that if the system does not produce the anticipated energy output, the residual energy costs from net-supplied electricity in addition to monthly solar equipment rental payments could result in paying more for solar energy on a leveled basis than for conventional electricity [35].

Incentives and regulations also vary from state to state, so it is important that municipal organizations in different regions assess solar power feasibility based on the local regularory environment. At the time, Arizona had several incentives and policies an impact on the economics of solar feasibility. The Federal Energy Policy Act of 2005 requires state utility commissions and large utility providers not regulated by the commission to assess their interconnectivity and net metering procedures [36]. Interconnection standards govern how an on-site solar generating system is allowed to physically connect to the grid [37]. Net metering is a billing arrangement that establishes the right of an owner to be compensated by the utility company for any net excess generation produced by the system; this is accomplished using a single, bi-directional meter. Net metering rules were adopted by the Arizona Corporation Commission in October 2008, and approved by the Attorney General in March [38]. Under Arizona rules, net metering is available to customers who generate electricity from solar or other renewable energy source. While the Arizona Corporation Commission does not regulate Salt River Project, the utility has established distributed generation interconnection guidelines and a net metering program for its customers [39, 40]. At the time of the Mesa project assessment, Salt River Project's EarthWise™ Solar Energy Program provides a one-time incentive of $1 per watt, up to a maximum of $30,000, for commercial systems up to 30 kW [41]. In exchange for the incentive, Salt River Project retains the solar renewable energy certificate associated with the system [42].

IDENTIFYING SUITABLE FACILITIES

Large-scale implementation of photovoltaic in an urban environment presents considerable challenges. To estimate the rooftop solar feasibility of fire stations and potential performance of PV systems, aerial images were analyzed to determine suitable space and then peak wattage and annual energy production factors were calculated using RETScreen® software. This was compared to an inventory of emissions and annual electrical consumption data for fire stations to select appropriate facilities for retrofit projects. The method applied a scheme created by Bryan et al. for estimating rooftop solar feasibility on an urban scale [43].

Aerial Images Analysis

Aerial images were acquired for fire department facilities using iMaps from the Mesa Planning Department. iMap uses ArcGIS (version 1.3) interactive applications over images from Google Earth; images date October 31, 2009. The iMap application has a robust set of tools and controls including search (by address, street intersection, parcel, or planning case), planning case drill down tool, parcel information, map layer control, Google street view, and measuring tool. Images of fire department facilities were located by entering the address into the search function. The images were clear enough to allow an initial visual assessment of roof type, rooftop equipment and penetrations, building orientation, and shading. Buildings were given a rating of one (1) through five (5), with a rating of 5 having the highest potential for photovoltaic installations and 1 having the lowest. Buildings rated a 4 or 5 were further assessed for available space. Rooftop areas suitable for photovoltaic were highlighted and measured using a measuring tool embedded in the software. Selecting the measuring tool opens a panel of options for using an interactive approach for measuring distance and area, or for determining the coordinate at a click location. Area measurements were reported in square feet (ft2). One feature enables the user to draw a polygon to measure area allowing for more accurate measurement of unusually shaped roof perimeters.

Estimating System Performance

The total solar power production capacity potential for each facility was estimated for eight different combinations of photovoltaic systems. Peak wattage (kW•p/ft2) and annual kW•h/ft2 production were calculated for thin film/amorphous with 8% efficiency, poly-crystalline silicone with 12% efficiency, and mono-crystalline with 16% efficiency, using three different mounting systems; horizontal laminate system, a ballasted system with 18o tilt, and a one way tracking system. These efficiency ratings and tilt angles ensured conservative power output estimates. A peak wattage factor and annual energy production factor were determined for all systems using RETScreen® software. Rooftop area suitable for photovoltaic installation, as determined in aerial image analysis, was multiplied by the factors in Table 1 to determine peak wattage and annual kW•h production for each fire department facility.

Table 1. Peak wattage and annual kWh production for different types of PV systems (based on [43])

	Fixed 0° Tilt	Fixed 18° Tilt	Single-Axis
Thin film/Amorphous			
Peak kWp/ft²	0.00666	0.00592	x
Annual kWp/ft²	11.583	11.208	x
Poly-crystalline			
Peak kWp/ft²	0.01008	0.00896	0.00896
Annual kWp/ft²	16.479	15.912	21.096
Mono-crystalline			
Peak kWp/ft²	0.01332	0.01184	0.01184
Annual kWp/ft²	21.78	21.024	27.88

Calculating Consumption and Emissions

In addition to an initial assessment of aerial images and calculations of potential system performance, consideration was also given to the electrical consumption of fire department facilities and their calculated pollutant emissions. Electrical consumption varies from facility to facility, and has impacts on photovoltaic system feasibility. All other factors being equal, it is preferable to invest in solar on buildings with higher electrical costs and peak demand loads. Both the annual power consumption (kW•h) and average electricity rate (₡/kW•h) were reported; these figures were calculated using fire department utility expense reports. Salt River Project [40] has a time-of-use pricing plan that applies different rates depending upon time of day and season (summer, summer peak, and winter): on peak ($0.1347 - $0.1667/kW•h), shoulder ($0.0979 - $0.1075/kW•h), and off-peak ($0.0531 - $0.0601/kW•h). By calculating the average electricity rate (₡/kW•h), this method takes into account which facilities have higher on-peak energy consumption, rather than just looking at higher overall energy use.

Pollutant output of fire department facilities was determined for pounds of emissions for sulphur dioxide (SO_2), oxides of nitrogen (NOX), particulate matter (PM10), carbon dioxide (CO_2), and volatile organic compounds. These pollutants are major contributors to global warming and poor air quality in the Phoenix region. Emissions can be adjusted based on the fuel used at the power generating station, including those utilizing western coal, eastern goal, natural gas, biomass, or oil. The majority of the city of Mesa is certified to be served by Salt River Project. Electrical power for the Mesa area is generated at the San Tan Generating Station, which is fueled by natural gas; thus, pollution calculations were made for natural gas [28, 44]. Because power plants are typically only 36% efficient, they produce large amounts of waste heat and require copious amounts of water for cooling (Texas State Energy Conservation Office n.d.), estimates of the gallons of water used for cooling were also calculated. Data were entered for the total annual pollutant output for each facility.

DISCUSSION

An audit of the energy consumption and pollution output of all fire department facilities quantifies the department's overall contribution to greenhouse gas and particle pollution in the region. Fire department facilities in the Salt River Project service area collectively used 3,153,249 kW•h of energy in fiscal year 09/10. This amount of energy produced 18.92 pounds of SO_2, 3,595 lbs. of NOX, 126.1 lbs. of PM10 emissions, 4,036,000 lbs. of CO_2, and 315.3 lbs. of volatile organic compounds. In providing this electricity to fire department facilities, the power plant used and estimated 2.365 million gallons of water for cooling. Power from Salt River Project provided to the City of Mesa is generated at the San Tan Power Generating Station in Gilbert, three miles south of the Mesa city limits. As a result, all of the pollutants emitted from this plant directly contribute to Mesa's air quality problems.

Several factors were used to evaluate which fire department facilities had the higher solar energy potential. These included an assessment of aerial images to identify usable rooftop space, power consumption and pollutant output, and calculations for system performance. Table 2 presents the data for rooftop area, power consumption, and electricity costs for five fire department facilities rated a 4 or 5 from an aerial image analysis. After an examination of aerial images, two facilities were given a rating of 5 for maximum available rooftop space with no shading and a sound roof structure. Fire Station 212 is a joint public safety facility that includes the Superstition Patrol Division. This facility has 17,818 square-feet (s.f.) under roof and an estimated usable rooftop area of 8,300 s.f. Also rated a 5, the Public Safety Training Academy has a 16,244 s.f. auditorium and classroom building with over 5,400 s.f. of usable rooftop area. Fire Station 206 is another large facility that includes a volunteer center with a total of 17,206 s.f. of space; however, 206 was rated a 4 because of its metal-clad roofing material and mostly east-west low-pitch orientation. Otherwise, the Station 206 has a maximum potential usable rooftop space of up to 7,100 s.f. Fire Stations 216 and 217 were rated a 4. Both are smaller facilities with a low-pitched roof with east-west orientation, and some shading issues from HVAC locations.

Table 2. Power consumption and costs, 2009/10

Facility	Square Foot	Usable Area	Rating	Energy Cost ($)	Annual kWh	₡/kWh
Station 206	17,264	7100	4	22,164.12	238,160	9.30
Station 212	17,818	8300	5	39,654.50	430,560	9.20
Station 216	9,012	3500	4	9,991.06	101,640	9.98
Station 217	10,284	3000	4	13,821.61	144,240	9.58
Training Academy	24,235	5800	5	53,321.16	518,960	10.28

Table 2 also highlights electricity consumption for these facilities. Fire Station 212 and the Public Safety Training Academy had the highest energy costs in FY09/10 at 430,560 kWh and 518,960 kWh, respectively. The

Academy also had the highest cost for energy at $53,321.16 and highest cost per kWh of 10.28₡/kWh. Surprisingly, Fire Station 212, while having the second highest cost for energy, $39,654.50, had the lowest cost per kWh at 9.2₡/kWh; this is likely due to less on-peak usage.

Table 3. Pollutant output, 2009/10

Facility	Ann kWh	SO_2	NO_X	PM_{10}	CO_2	VOC	Water
Station 206	238,160	1.429	271.5	9.526	304800	23.82	178600
Station 212	430,560	2.583	490.8	17.220	551100	43.06	322900
Station 216	101,640	0.6098	115.9	4.066	130100	10.16	76230
Station 217	144,240	0.8654	164.4	5.770	184.600	14.42	108200
Training Academy	518,960	3.114	591.6	20.760	664,300	51.9	389200
All MFD Facilities	3,153,249	18.92	3,595	126.100	4,036,000	315.3	2,365,000

Source: [45].

Table 3 documents pollutant output. Not surprisingly, facilities with higher usage produced more greenhouse gas and particle pollutants. It should be noted, however, that this might not be the case for every jurisdiction. Large urban areas or countywide fire departments may receive power from two or more suppliers. These suppliers may produce energy from different sources of fossil fuel such as western coal, eastern coal, natural gas, or nuclear; some may use renewable energy sources such as concentrated solar photovoltaic, wind, hydroelectric, or geothermal. The pollutant output of each fire department facility will depend upon who provides the power and what sources of energy are being used.

Finally, RETScreen® modeling found that solar PV system performance under Phoenix-area conditions for horizontal and slightly tilted systems ranged from peak solar output of 0.00666 kW/ft2 to 0.01184 kW/ft2. From these estimates, calculations for peak wattage and annual kW•h production were made for three different PV types and three mounting systems.

Table 4. Estimates of solar energy output for Fire Station 212

Fire Station	212
Address	2430 S. Ellsworth Mesa, AZ 85208
Facility Area (s.f.)	17,818
Est. solar ready roof area (s.f.)	5400
Roof type	Flat
Annual Energy Consumption (kWh)	430,560
Annual electricity cost ($)	39,654.50
Avg. Elec. Rate ₵/kWh	9.20
Solar suitability rating (1-5)	5
Peak kWp A-Si 0° tilt Annual kWh A-Si 0° tilt	55.28 96138.9
Peak kWp A-Si 18° tilt Annual kWh A-Si 18° tilt	49.14 93026.4
Peak kWp P-Si 0° tilt Annual kWh P-Si 0° tilt	83.66 136775.7
Peak kWp P-Si 18° tilt Annual kWh P-Si 18° tilt	74.37 132069.6
Peak kWp P-Si single axis Annual kWh P-Si single axis	74.37 175096.8
Peak kWp M-Si 0° tilt Annual kWh M-Si 0° tilt	110.56 180774
Peak kWp M-Si 18° tilt Annual kWh M-Si 18° tilt	98.27 174499.2
Peak kWp M-Si single axis Annual kWh M-Si single axis	98.27 231404

Notes. Large, unobstructed roof area over police patrol division. Very limited shading issues or roof penetrations. Newer facility. Mechanicals concentrated off to the side.

Figure 1. Side (left) and aerial image (right) of Fire Station 212.

With the largest estimated available rooftop area of 8,300 s.f., Fire Station 212 (depicted on Figure 1) recorded the highest solar energy potential ranging from 93,026.4 annual kW•h for thin-film/amorphous at a fixed 18o angle, to 231,404 for mono-crystalline panels in a single-axis configuration. This was followed by Fire Station 206 (7,100 s.f.; 79,576.8 to 197,948 annual kW•h) and the Public Safety Training Academy (5,800 s.f.; 65,006.4 to 161,704 annual kWh). It is worth noting that none of the facilities evaluated reached net-zero energy use [3]; that is, it produced as much energy as it uses. Fire Station 216 was close. In FY09/10 the station used 101,640 kWh of energy. The highest estimated annual kW•h of output for solar photovoltaic was 97,580 kW•h, or 96% of the energy needs. The public safety training area, on the other hand, used 518,960 kW•h in FY09/10, and had a maximum estimated solar energy output of 150,552 annual kW•h based on 5,800 s.f. of available rooftop space; only 29% of the facility's energy needs. Table 4 provides an example of the calculation outcomes.

CONCLUSION

Converting buildings with higher solar energy potential to photovoltaic power has many benefits for the city of Mesa and the fire department. The economic benefits include a reduction in the operating costs of fire department facilities through an offset in grid-supplied electricity use; the reduction in fossil fuel consumption and lowering of greenhouse gas and particulate matter emissions provides an environmental benefit; and there is a social benefit through public education and awareness of the positive impacts of renewable energy sources. However, cities must guard against an unrealistic expectation that solar power has the capability to wholly solve budget shortfalls or regional air quality issues.

Solar is not a panacea for fiscal problems. This study showed that, at best, converting the two sites with the highest photovoltaic feasibility would account for an operating expense reduction of $36,574.21 for electricity consumption, assuming these buildings reach net-zero energy consumption

[3], which is unlikely. The net savings from a solar conversion would depend upon electricity consumption, future energy prices, and contractual arrangements set forth in a power purchase agreement.

Similarly, the contribution of pollutants produced by fire department facilities is minute in relation to the overall output of the San Tan Generating Station. At full capacity, San Tan is rated to generate 1,225 MW of electricity [44]. In fiscal year 09/10, fire department facilities in the Salt River Project service area used 3153.25 MW•h, or an estimated 0.357 MW of energy, a mere fraction (2.914e-4) of the electricity generated.

Even if the economic return is cost neutral, and the reduction of greenhouse is minimal, there is a value to the public image and education opportunity afforded to the fire service for installing renewable energy systems. Solar PV on fire stations not only serves to educate the public about alternative energy sources, but also the negative public health and safety externalities of conventional energy generation. Ultimately, however, environmental degradation respects no political boundaries; therefore, regional approaches to resolve environmental issues that cross-jurisdictional boundary should be encouraged.

REFERENCES

[1] Gamble, J. L., Ebi, K. L., Grambsch, A. E., Sussman, F. G., and Wilbanks, T. J. (2008). *Analyses of the effects of global change on human health and welfare and human systems: Final report, synthesis and assessment product.* Report by the US climate change science program and the subcommittee on global change research. Washington, DC, US.

[2] National League of Cities. (2011). *National municipal policy. Washington, DC: National League of Cities.* http://www.nlc.org/influence-federal-policy/resources/national-municipal-policy.

[3] Torcellini, P., Pless, S., Deru, M., & Crawley, D. (2006). Zero energy buildings: a critical look at the definition (No. NREL/CP-550-39833).

National Renewable Energy Lab. (NREL), Golden, CO (United States).

[4] US Environmental Protection Agency. (2020). *Particulate matter (PM) pollution.* Last modified May 8, 2020. https://www.epa.gov/pm-pollution.

[5] Tebaldi, C., Adams-Smith, D., and Heller, N. (2013). The heat is on: US temperature trends. *Climate Central.*

[6] City of Mesa. (2010). *Executive budget plan City of Mesa*, Arizona for the fiscal year 2010/2011. Retrieved from: http://www.mesaaz.gov/budget/Documents/FY_10_11/City_of_Mesa_AZ_FY_2010-2011_Executive_Budget_Plan.pdf.

[7] Maricopa County. (2011). *Inversion. Air Quality Department.* Accessed April 15, 2011. http://azdeq.gov/what-inversion-and-how-does-it-affect-air-quality (2011).

[8] Eastwood, S. (2006). *Brown air in the valley: Air pollution creates a brown cloud over Phoenix.* Accessed April 21, 2011 from http://phoenix.about.com/od/ weather/a/browncloud.htm.

[9] American Lung Association. (2011). *State of the air 2011.* Phoenix, AZ. Retrieved by the author from www.stateoftheair.org.

[10] Fierro, M. (1999). Ozone Health Effect. *First Revision*, JB.

[11] US Environmental Protection Agency. (2004). *Air Quality Criteria for Particulate Matter.*

[12] Arizona Department of Environmental Quality. (2011). ADEQ, Maricopa County choose to improve dust control plan: ADEQ withdraws earlier plan to U.S. *EPA*. Accessed March 30, 2011. http://lib.trinity.edu/research/citing/apaelectronicsources.pdf.

[13] Mar, T. F., Norris, G. A., Koenig, J. Q., & Larson, T. V. (2000). Associations between air pollution and mortality in Phoenix, 1995-1997. *Environmental health perspectives*, 108(4), 347-353.

[14] Grineski, S., Bolin, B., and Boone, C. (2007). Criteria air pollution and marginalized populations: environmental inequity in metropolitan Phoenix, Arizona. *Social Science Quarterly* (88 no. 2), 535-554.

[15] Parry, M., Parry, M. L., Canziani, O., Palutikof, J., Van der Linden, P., & Hanson, C. (Eds.). (2007). *Climate change 2007-impacts,*

adaptation and vulnerability: Working group II contribution to the fourth assessment report of the IPCC (Vol. 4). Cambridge University Press.

[16] Arizona Climate Change Advisory Group. (2006). *Climate change action plan*. Retrieved by the author from www.azclimatechange.gov/docs.html.

[17] Bailie, A., Lazarus, M., Peterson, T., Hausker, K., Kuch, P., Williams, E., Colburn, K. and Roe, S. (2005). *Draft Arizona Greenhouse Gas Inventory and Reference Case Projections 1990–2020.* Arizona Climate Change Advisory Group.

[18] Owens, S. (2009). Climate Change Action in Arizona. *UCLA Journal of Environmental Law & Policy* 27, 317.

[19] Westerling, A. L., Hidalgo, H. G., Cayan, D. R., & Swetnam, T. W. (2006). Warming and earlier spring increase western US forest wildfire activity. *Science*, 313(5789), 940-943.

[20] Bonnicksen, T. M. (2008). Greenhouse gas emissions from four California wildfires: Opportunities to prevent and reverse environmental and climate impacts. *FCEM Report 2.*

[21] Arizona Department of Environmental Quality. (2011). *Fires, smoke, and your health*. Accessed March 29, 2011. http://www.azdeq.gov/environ/air/smoke/download/fire_smoke_your_health_brochure.pdf.

[22] Considine, T. J., & McLaren, D. (2008). *Powering Arizona: Choices and Trade-Offs for Electricity Policy*. The Communications Institute, Arizona State University: Tempe, AZ, 1-34.

[23] Valley Forward Association. (2011). *Arizona's guide to electric energy choices.* Phoenix, AZ: Valley Forward Association. http://www.valleyforward.org/issues/22/.

[24] MacIntosh, D. L., Spengler, J. D., Staudt, J. E., & Bachmann, J. (2011). Emission of hazardous air pollutants from coal-fired power plants. *J Environ Sci Health*, 10, 1-39.

[25] Kobayaski, T. & Hung, L. (2008). Global warning and the fire service: A local view from Asia. *Fire & Rescue* (69), 38-40.

[26] Betsill, M. M. (2001). Mitigating climate change in US cities: opportunities and obstacles. *Local environment*, 6(4), 393-406.

[27] US Conference of Mayors. (2009). *1,000th mayor – Mesa, AZ Mayor Scott Smith signs the U.S. Conference of Mayors climate protection agreement.*

[28] Mesa Fire Department. (2006). *Strategic Plan 2006*. Mesa, AZ: City of Mesa, 2006. http://mesaaz.gov/fire/Technical/pdf/2006FIRESTRATEGICPLAN.pdf

[29] City of Mesa. (2002) *City of Mesa General Plan*. Retrieved from : http://www.mesaaz.gov/planning/PDF/GeneralPlan/MesaGeneralPlan.pdf.

[30] City of Mesa. (2010). "*2040 General Plan*." Mesa, AZ: City of Mesa, 2010. https://www.mesaaz.gov/home/showdocument?id=12298.

[31] City of Mesa. (2009). *Executive budget plan City of Mesa, Arizona for the fiscal year 2009/2010*. Retrieved from: http://www.mesaaz.gov/budget/Documents/FY_09_10/FY_2009-2010_City_of_Mesa_Executive_Budget_Plan.pdf.

[32] Borenstein, S. (2008). *The market value and cost of solar photovoltaic electricity production.*

[33] Barbose, G. L., Wiser, R., Peterman, C., and Darghouth, N. (2009). *Tracking the Sun II: The Installed Cost of Photovoltaics in the US from 1998-2008*. Report No. LBNL-2674E. Lawrence Berkeley National Lab. (LBNL), Berkeley, CA (United States).

[34] Database of State Incentives for Renewables & Efficiency. (2011). *Clean renewable energy bonds* (*CREBs*). Accessed March 16, 2011. http://www.dsireusa.org/incentives/index.cfm?State=US&ee=1&re=1

[35] Kollins, K. (2009). *Solar PV Financing: Potential Legal Challenges to the Third Party PPA Model* (Doctoral dissertation, Duke University).

[36] Keyes, J., Fox, K., and Wiedman J. (2009). *Net Metering and Interconnection Procedures-Incorporating Best Practices. Report No. IREC401-002*. Interstate Renewable Energy Council, 2009.

[37] Database of State Incentives for Renewables & Efficiency (2011). *Qualified energy conservation bonds* (*QECB*). Accessed March 16, 2011, http://www.dsireusa.org/incentives/index.cfm?State=US&ee=1&re=1.

[38] Database of State Incentives for Renewables & Efficiency. (2011). *Arizona – Net metering*. Accessed March 17, 2011. http://www.dsireusa.org/incentives/index.cfm?re=1&ee=1&spv=0&st=0&srp=1&state=AZ.

[39] Database of State Incentives for Renewables & Efficiency. (2011). *SRP – EarthWise solar incentive program*. Accessed March 17, 2011. http://www.dsireusa.org/incentives/index.cfm?re=1&ee=1&spv=0&st=0&srp=1&state=AZRooftop Solar Photovoltaic 64

[40] Salt River Project. (2011). *Business time-of-use plan (E-32)*. Accessed April 2, 2011. http://www.srpnet.com/prices/business/tou.aspx.

[41] Salt River Project. (2011). *Solar electricity for your business from SRP EarthWise Solar Energy*. Accessed March 31, 2011. http://www.srpnet.com/environment/earthwise/solarbiz/solarelectric.aspx.

[42] Database of State Incentives for Renewables & Efficiency. (2011). *SRP – Net metering*. Accessed March 17, 2011. http://www.dsireusa.org/incentives/index.cfm?re=1&ee=1&spv=0&st=0&srp=1&state=AZ.

[43] Bryan, H., Rallapalli, H., Rasmussen, P. and Fowles, G. (2010). *Methodology for estimating the rooftop solar feasibility on an urban scale*. In Conference proceedings of the SOLAR 2010.

[44] Salt River Project. (2011). *San Tan generating station*. Accessed April 2, 2011 from http://www.srpnet.com/about/stations/santan.aspx

[45] Texas State Energy Conservation Office. (2011). *Electric power pollution calculator*. Accessed March 21, 2011, http://www.infinitepower.org/calc_pollution.htm.

BIOGRAPHICAL SKETCH

Scott Somers

Affiliation: Watts College of Public Service and Community Solutions, Arizona State University

Education: Ph.D. Arizona State University

Business Address: 411 N. Central Ave #750, Phoenix, AZ 85004

Research and Professional Experience:

Scott Somers is a professor of practice with the College of Public Service and Community Solutions at Arizona State University, bringing over two decades of public safety and emergency management experience. He also serves as a senior fellow with the Goldwater Center for Cyber and Homeland Security. Previously, he held positions with the Phoenix Fire Department and FEMA Urban Search and Rescue AZ-TF-1 team. In fact, Somers was one of 80 AZ-TF-1 members who conducted search and rescue missions during Hurricane Harvey and Hurricane Irma in August and September 2017. His research interests include urban resilience, environmental risk and adaptation, terrorism and political violence, public safety technology and mobile integrated healthcare.

Professional Appointments:

Somers has also held numerous elected and appointed leadership roles, most recently as a city council member in Mesa, Arizona. Additionally, he has served as chair of the National League of Cities Public Safety and Crime Prevention, on the National EMS Advisory Council, National Homeland Security Consortium and the FirstNet PSAC & SAFECOM Executive Committee.

Honors:

Recipient of the 2010 Louis Brownlow Award for the best peer-reviewed article in Public Administration Review. Two-time recipient of the Outstanding Applied Research Award, Executive Fire Officer Program, National Fire Academy 2011-12.

The present chapter was written based on Scott Somers' USFA National Fire Academy executive fire officer program study, Rooftop Solar

Photovoltaic Feasibility Study for the Mesa Fire Department. The project received an A. Don Manno Award for best applied research.

Publications from the Last 3 Years:

1. Somers, S. (2020). How terrorists leverage climate change. In *Resilience Matters: Action in an Age of Uncertainty*. Laurie Mazur (Ed.), Washington, DC: Island Press, 2020, 21-24.
2. Somers, S. (2017). Difficulty identifying universal high-efficacy locations for public access defibrillators/*International Journal of Emergency Services*, 6(2), 84-98.

About the Editors

Axel Cablé, PhD

R&D Manager, EXEC – CEO, IBS-Conseil

Independent Expert accredited by the French Ministry of Higher Education, Research and Innovation

Axel Cablé is an independent expert accredited by the French Ministry of Higher Education, Research and Innovation for R&D operations in the fields of Fire and Industrial Safety, Energy Efficient Buildings and Composite Materials. M. Cablé holds a PhD in Fluid Mechanics and an Engineering degree in Energy Sciences. He has 11 years of experience in leading research institutes in France and Norway (UMR CNRS, CEA, INSA, SINTEF), and is specialized in the development, numerical modelling, experimental testing, and certification process of innovative technologies. He is currently the R&D Manager of the SME EXEC (Paris Area, France) which developed and patented the novel Vacuum Sprinkler system, and CEO of the company IBS-Conseil (Lyon Area, France) which specializes in Regional, National and European funding of R&D projects and innovative technologies. M. Cablé is also involved as external expert in Fire Safety for the National Institute of Applied Sciences (France) in the context of several research projects related to the fire protection of cultural heritage and critical infrastructures.

Brady Manescau, PhD
Associate Professor
National Institute of Applied Sciences, France

Brady Manescau is Associate Professor in Fire and Industrial Safety Engineering at the National Institute of Applied Sciences, Orléans University (Bourges, France) and Researcher in the Fluids, Energy, Combustion, and Propulsion (FECP) Department of the PRISME EA 4229 laboratory. M. Manescau holds a PhD in Fire Sciences and an Engineering degree in Energy and Fire Safety. His fields of research include the Fire Behavior of Composite Materials, Smoke Spreading, Reduced Scale and Full Scale accredited Fire Testing, Computational Fluid Dynamics and Fire Modelling, amounting to 10 years of experience in forefront research laboratories in France (UPR CNRS, PPRIME, INSA), with applications notably in the Defense and Aeronautics industries.

INDEX

A

B

C

D

E

R

S

T

U

V

W

X

Y